Gianluca Cama

Impedance measurement techniques for the analysis of cell behavior

Gianluca Cama

Impedance measurement techniques for the analysis of cell behavior

Disposable, micro-fluidic biosensor array for online analysis of adherent cells activity

Südwestdeutscher Verlag für Hochschulschriften

Impressum/Imprint (nur für Deutschland/only for Germany)
Bibliografische Information der Deutschen Nationalbibliothek: Die Deutsche Nationalbibliothek verzeichnet diese Publikation in der Deutschen Nationalbibliografie; detaillierte bibliografische Daten sind im Internet über http://dnb.d-nb.de abrufbar.
Alle in diesem Buch genannten Marken und Produktnamen unterliegen warenzeichen-, marken- oder patentrechtlichem Schutz bzw. sind Warenzeichen oder eingetragene Warenzeichen der jeweiligen Inhaber. Die Wiedergabe von Marken, Produktnamen, Gebrauchsnamen, Handelsnamen, Warenbezeichnungen u.s.w. in diesem Werk berechtigt auch ohne besondere Kennzeichnung nicht zu der Annahme, dass solche Namen im Sinne der Warenzeichen- und Markenschutzgesetzgebung als frei zu betrachten wären und daher von jedermann benutzt werden dürften.

Coverbild: www.ingimage.com

Verlag: Südwestdeutscher Verlag für Hochschulschriften GmbH & Co. KG
Dudweiler Landstr. 99, 66123 Saarbrücken, Deutschland
Telefon +49 681 37 20 271-1, Telefax +49 681 37 20 271-0
Email: info@svh-verlag.de

Approved by: Magdeburg, OVGU, Diss., 2011

Herstellung in Deutschland:
Schaltungsdienst Lange o.H.G., Berlin
Books on Demand GmbH, Norderstedt
Reha GmbH, Saarbrücken
Amazon Distribution GmbH, Leipzig
ISBN: 978-3-8381-2958-7

Imprint (only for USA, GB)
Bibliographic information published by the Deutsche Nationalbibliothek: The Deutsche Nationalbibliothek lists this publication in the Deutsche Nationalbibliografie; detailed bibliographic data are available in the Internet at http://dnb.d-nb.de.
Any brand names and product names mentioned in this book are subject to trademark, brand or patent protection and are trademarks or registered trademarks of their respective holders. The use of brand names, product names, common names, trade names, product descriptions etc. even without a particular marking in this works is in no way to be construed to mean that such names may be regarded as unrestricted in respect of trademark and brand protection legislation and could thus be used by anyone.

Cover image: www.ingimage.com

Publisher: Südwestdeutscher Verlag für Hochschulschriften GmbH & Co. KG
Dudweiler Landstr. 99, 66123 Saarbrücken, Germany
Phone +49 681 37 20 271-1, Fax +49 681 37 20 271-0
Email: info@svh-verlag.de

Printed in the U.S.A.
Printed in the U.K. by (see last page)
ISBN: 978-3-8381-2958-7

Copyright © 2011 by the author and Südwestdeutscher Verlag für Hochschulschriften GmbH & Co. KG and licensors
All rights reserved. Saarbrücken 2011

AKNOWLEDGMENT

The present work arises from my employment as a scientific assistant at the Institute of Experimental Internal Medicine at the Medical Faculty of Magdeburg in collaboration with the Institute of Micro and Sensor Systems at the University of Magdeburg. The necessary founding was provided by the EU Marie Curie Research Training Network "On-chip cell handling and analysis", reference: MRTN-CT-2006-035854.

First of all I want to thank my doctoral advisor Prof. Dr. Peter Hauptmann for drawing my attention to the field of quartz crystal resonators as well as giving me the opportunity to carry out my research work with all the necessary support in each aspect.

I express my gratefulness to Prof. Dr. Michael Naumann who gave me the opportunity to work for a medical institution contributing to my scientific improvement in a multidisciplinary area such as biomedical engineering.

I also thank Prof. Winnie E. Svendsen and her research group making my stay at the Technical University of Denmark within the Cellcheck project a valuable and pleasant experience.

Further I would like to thank all my colleagues at the Institute of Micro and Sensor Systems for their kind assistance.

Finally, the cherished gratitude of my heart goes to my wife Nina. Writing this thesis would not have been possible without her patience and support.

ABSTRACT

In the present work, a novel micro-fluidic biosensor array for fast online adherent cell monolayer (e.g. living epithelial cells) analysis during cell proliferation and stimulation was developed. The new device combines 16 quartz crystal resonators and 16 impedimetric sensors to acquire complementary sets of measurement data, namely the acoustic shear response and the transepithelial impedance of the cell layer. Sensor responses are analysed by means of impedance spectroscopy in the MHz range. Time-lapse light microscopy through transparent microelectrodes was employed for visual characterisation of the cell monolayers. To allow parallelized cell cultivation the new biosensor array consists of 4 bioreactor units and a flow distribution network embedded within the same device. Each bioreactor unit houses 4 quartz crystal resonators and 4 impedimetric sensors for the simultaneous analysis of four cell populations. In order to achieve the parallel cultivation of different cell populations the flow distribution network was designed with symmetric structure. This enables the equal partitioning of the cell suspension and the media providing the same initial cell concentration and identical conditions during cell growth and stimulation among all the cell populations cultivated within the biosensor array. Moreover, the array was designed to be operated in flow-through and overpressure regime. An external flow injection system provides automated and parallelized media feed as well as the control of the overpressure regime. This approach enables well defined proliferation and stimulation conditions. Besides, it prevents the accumulation of contaminants. Injection molding technology was chosen for the cheap mass production of the microfluidic array so that disposable parts made of biocompatible polymer could be fabricated. Thin-film deposition techniques were applied for the sensors fabrication. New dedicated sensor interface electronics including a multiplexer for all the 32 sensors were developed to allow fast and parallelized spectra acquisition with a miniaturized impedance analyser. To prove the assumption of equal flow circulation within the symmetric micro-channel network and support the hypothesis of identical cultivation conditions for the cells living above the sensors, the influence of fabrication tolerances on the flow regime has been simulated. As well, the shear stress on the adherent cell layer due to the flowing media was characterized. Furthermore, the injection molding process was simulated in order to optimize the mold geometry and minimize the shrinkage and the warpage of the parts as well as to ensure an even resident time of the melt in the mold. Madin-Darby Canine Kidney cells were cultivated in the biosensor array. During experiments, the cell behaviour during cell proliferation and stimulation was analyzed online. It is believed that in the future, the new biosensor array can be successfully employed as a tool supporting standard techniques employed in molecular cell biology for the study of that complex system of communication that governs basic cellular activities and coordinates cell actions.

ZUSAMMENFASSUNG

In der vorliegenden Dissertation wurde ein neues mikrofluidisches Biosensor-Array zur online Analyse adhärenter Zellkulturen, z.b. Epithelzellen, entwickelt. Das Sensorsystem beinhaltet 16 Quarzkristallresonatoren und 16 impedimetrische Sensoren zur parallelisierten Analyse adhärenter Zellkulturen hinsichtlich akustischer Schereigenschaften und der transepithalen Impedanz. Auf diese Weise kann das zelluläre Verhalten anhand zweier unabhängiger Sensorprinzipien überwacht werden. Auf Basis des Verfahrens der elektrischen Impedanzspektroskopie im Bereich MHz werden die relevanten Sensorparameter bestimmt. Das Array kann u.a. für die Analyse der Proliferation und Stimulation von Zellen eingesetzt werden. Dabei sollen die Messinformationen in Kombination mit etablierten Verfahren der molekularen Zellbiologie neue Einblicke in zelluläre Signal- und Regulationsmechanismen geben. Transparente Mikroelektroden ermöglichen die parallele Aufnahme von Lichtmikroskopbildern der Sensoroberflächen. Ein symmetrisches Mikrofluidkanalnetzwerk gestattet eine homogene Bedeckung aller Sensoren mit Zellen bzw. einem vergleichbaren Konfluenzgrad. Weiterhin wird eine definierte Dosierung der Zellkulturmedien und der Stimulanzien im Überdruck- und Durchflussregime in Verbindung mit einem externen Fließinjektionssystem gewährleistet. Letzteres verhindern eine Akkumulation von Kontaminationen und stellt konstante Wachstums- und Stimulationsbedingungen sicher. Im Hinblick auf die Einwicklung eines kostengünstigen Analysesystems wurden die wesentlichen mikrofluidischen Strukturen in Spritzgusstechnologie hergestellt. Die Spritzgusstechnologie auf Basis biokompatibler Kunststoffe gestattet die Massenproduktion kostengünstiger Einwegchips. Die Fertigung der Sensoren erfolgte mit Hilfe der Dünnschichttechnologie. Für die Impedanzspektrenmessung wird ein miniaturisierter Impedanzanalysator eingesetzt. Eine neu entwickelte Sensorinterfaceelektronik ermöglicht eine schnelle und präzise Messung der Impedanzspektren aller 32 Sensoren. Für die Optimierung der mikrofluidischen Strukturen wurde ein Simulationsmodel auf Basis von Comsol Multiphysics entwickelt. Das Modell wurde für die Analyse der Fließgeschwindigkeitsverteilung und des Einflusses von Scherkräften auf die Zellkultur einsetzt. Weiterhin wurde der Einfluss herstellungsbedingter Fertigungstoleranzen auf die Mikrokanalgeometrie untersucht. Auf Basis der Simulationsergebnisse wurde die Struktur des Sensorsystems festgelegt und entsprechend für die Messung im Überdruck- und Durchflussregime ausgelegt. Ein weiteres Simulationsmodell auf Basis von MoldFlow wurde für die Optimierung des Spritzgussprozess entwickelt. Besondere Aufmerksamkeit galt der Minimierung von Schrumpfung und einer gleichmäßigen Verweilzeitverteilung der Schmelze in der Spritzgussform. Das Biosensorarray konnte erfolgreich für die Analyse der Proliferation und Stimulation von MDCK-II Zellen eingesetzt werden.

SYMBOLS

A	area
A_{EL}	electrode effective area
C	capacitance
C_0	static capacitance
C_{ext}	external capacitance
C'_q	quartz motional capacitance
$C_{1,2}$	model capacitances according to the Maxwell-Wagner effect
C_G	Gouy-Chapman double layer capacitance
C_H	Helmholtz double layer capacitance
C_I	total double layer interfacial capacitance
C_W	Warburg capacitance
C_S	low frequency (static) capacitance
C_∞	high frequency capacitance
C_g	parasitic capacitance to ground
C_g	parasitic capacitance introduced by the circuitry
c_{q0}	piezoelectrically stiffened elastic stiffness constant
D	polarisation
D_0	"static" (low frequency) polarisation
D_∞	"instantaneous" (high frequency) polarisation
d	thickness
d_L	layer thickness
d_{OHP}	outer Helmholtz plane thickness
d_q	quartz thickness
E	electric field
f_c	relaxation characteristic frequency
f_0	mechanical resonance frequency
f_{res}	series resonance frequency at the conductance maximum of a loaded quartz

Symbols

$f_{res,0}$	series resonance frequency at the conductance maximum of a bare quartz
Δf, Δf_{liq}, Δf_{rig}	resonance frequency shift
Δf_{10}	frequency shift after 10 hours of cultivation
G_{el}	electrical conductance
G_L	load complex shear modulus
G_L'	load shear storage modulus
G_L''	load shear loss modulus
G_{max}, $G_{max,0}$	electrical conductance maximum
G_q	quartz complex shear modulus
G_q'	quartz shear storage modulus
G_q''	quartz shear loss modulus
k	Boltzmann constant
k_q	complex electromechanical coupling coefficient, AT-cut quartz
I	ionic strength of the electrolyte
i, j	imaginary unit
l	lenght
L_D	Debye lenght
L_q	motional inductance
L_{liq}	liquid loading inductance
L_{rig}	rigid loading inductance
$m_{1,2}$	constant of proportionality
$n_{1,2}$	constant of proportionality
N_A	Avogadro number
q	elementary charge
R	electric resistance
$R_{1,2}$	model resistances according to the Maxwell-Wagner effect
R_q	quartz motional resistance
R_{liq}	liquid loading resistance
R_0	low frequency (static) resistance
R_∞	high frequency resistance
R_S	spreading resistance

Symbols

R_W	Warburg resistance
R_{LOAD}	load resistor
r	radius
T	absolute temperature in kelvin
ΔR, ΔR_{liq}, ΔR_{rig}	resistance change
V	voltage
V_0	potential at the electrode
V_t	thermal voltage
v_q	acoustic wave velocity within the quartz
x	distance from the electrode
Z_{el}	electric impedance
Z_L	acoustic load impedance (effective)
Z_{liq}	acoustic load impedance for the non-gravimetric regime
Z_m	motional impedance of the quartz
Z_m^0	unperturbed QCR motional impedance
Z_m^L	load motional impedance
Z_{cq}	characteristic acoustic impedance of the quartz
Z_{rig}	acoustic load impedance for the gravimetric regime
Z_R	reference resistor
α	electric field angle
α_q	quartz acoustic phase shift
δ	penetration depth
η	viscosity
η_q	viscosity of quartz
η_{liq}	viscosity of a liquid
ρ_e	electrical resistivity
ρ	density
ρ_q	quartz density
ρ_L	density of a rigid layer
ρ_{liq}	density of a liquid
$\bar{\varepsilon}$	complex permittivity
ε_0	permittivity of vacuum
ε_r	relative permittivity of the material

Symbols

ε_S	static permittivity
ε_∞	high frequency permittivity
ε_{22}	dielectric permittivity of the quartz crystal
σ_S	low frequency conductivity
σ_∞	high frequency conductivity
τ	relaxation time
τ_k	kinetics time constant
$\omega = 2\pi f$	angular frequency
ω_0	mechanical angular frequency
ω_s	series resonance angular frequency

TABLE OF CONTENTS

AKNOWLEDGMENT I

ABSTRACT II

ZUSAMMENFASSUNG III

SYMBOLS IV

1. INTRODUCTION 1
   1.1. Motivation and state of the art 1
   1.2. Objectives 3
2. CELLULAR FUNCTION AND RESPONSE 5
   2.1. Cellular morphology 5
   2.2. The eukaryotic Cytoskeleton 7
   2.3. Cellular Adhesion 8
   2.4. Cellular Motility 9
   2.5. Membrane Impedance 10
3. FUNDAMENTALS 12
   3.1. The Quartz Crystal Resonator 12
      3.1.1. Model of the sensor 13
         3.1.1.1. The thickness shear mode and acoustic load concept 13
         3.1.1.2. The Transmission-Line Model 15
         3.1.1.3. The modified Butterworth van Dyke Model 17
         3.1.1.4. The load motional impedance 19
      3.1.2. Impedance analysis of the shear wave resonator 21
   3.2. Microelectrode impedance spectroscopy 22
      3.2.1. The metal-electrolyte interface 23
         3.2.1.1. Interface capacitance 23
         3.2.1.2. Solution resistance 25
      3.2.2. Cell layer electric model 26

Table of contents

        3.2.2.1. Relaxation and dispersion 26

        3.2.2.2. Maxwell-Wagner effects 27

   3.3. Injection molding technology 29

      3.3.1. Mold design and fabrication 29

      3.3.2. Influence of processing parameters 31

4. DESIGN OF DEVICES AND MOLD FABRICATION 32

   4.1. Resin selection 32

   4.2. Microfluidic network design and simulation 32

      4.1.1. COMSOL simulations 34

   4.3. Molding process simulation and optimization 36

      4.3.1. Gate location 37

      4.3.2. Optimisation of the gate geometry 38

      4.3.3. Optimization of process parameters 39

5. FABRICATION OF SENSORS AND MEASUREMENT TECHNOLOGY 42

   5.1. Layout and fabrication of QCRs' electrodes 42

   5.2. Layout and fabrication of transparent ITO microelectrodes 44

      5.2.1. Laser ablation of ITO layers 45

      5.2.2. Wet etching procedure and mask fabrication 46

      5.2.3. Comparison of laser ablation and wet etching 47

   5.3. The measurement unit 47

   5.4. Measurement electronics 49

      5.4.1. Impedance analysis 50

   5.5. Characterization of the sensor array 51

      5.5.1. Signal drift, precision and accuracy 54

6. SENSOR SYSTEM ARCHITECTURE 55

   6.1. Overview of the measurement system 55

   6.2. Multi-position time lapse microscopy 56

   6.3. Flow injection system 57

Table of contents

	6.3.1. Integrated shut off valves	58
7.	DESIGN OF EXPERIMENTS	61
7.1.	Response to liquid flow	61
7.2.	Changes in the composition of cell culture media	62
7.3.	Cell seeding procedure	64
7.4.	Stimulation of cells	64
7.5.	Optimised experimental procedure	65
8.	CELL CULTURE EXPERIMENTS	67
8.1.	Cell proliferation and cell detachment	67
	8.1.1. QCRs response	68
	8.1.2. ITO microelectrodes response	70
8.2.	Cell starvation	74
8.3.	Impact of the flow regime on cell proliferation	75
8.4.	Impact of cell distribution on proliferation kinetics	77
8.5.	Cell stimulation with Hepatocyte growth factor	81
	8.5.1. QCRs response	83
	8.5.2. ITO microelectrodes response	84
8.6.	Discussion of results	85
9.	CONCLUSIONS AND OUTLOOK	88
REFERENCES		92

Table of contents

1. INTRODUCTION

1.1. Motivation and state of the art

Cell adhesion and motility of eukaryotic cells is of great importance in several biological processes, e.g. maintenance of tissue and wound healing as well as embryonic development. As initial step, cell attachment is a central prerequisite for cell movement, and focal adhesion complexes are developed during this process [1]. Deregulation of these processes is associated with many diseases like pathogen infections and metastatic cancer [2, 3]. Thus, online analysis of cell adhesion and motility could gain new insights in several biological processes. Different methods are available for the quantification of cell adhesion to a surface. These embrace direct counting techniques such as labeling of cells after fixing or enzymatic detachment [75] as well as indirect counting techniques like spectrofluorimetry of cell stain after cell lysis [76, 77]. Moreover, adhesion forces can be investigated through the combination of fluorescence intensity measurements with detachment in a centrifugation assay [78]. Cell area and density can be measured by means of a variety of optical methods [74, 77, 79]. Due to the interaction of cells with the substrate, Quartz Crystal Resonator (QCR) based techniques are excellent and versatile candidates to be employed as in vitro methods for real-time characterization of cell adhesion and motility processes with no need for destructive interventions. Thanks to its high mass sensitivity the Quartz Crystal Microbalance (QCM) has been widely employed in thin film deposition applications, in surface science for adsorption studies and in analytical chemistry [80] as well as for the quantitative characterization of viscoelastic coatings [44, 81] and gas detection [82]. Moreover, it has also demonstrated to be a powerful tool for measuring cell spreading and adhesion [5, 7, 83 ,84] along with changes in the cell cytoskeleton [85]. In order to evaluate the viscoelastic properties of the adherent cell layer attached on the surface of the sensors various parameters have been recorded together with the resonance frequency. The motional resistance [6, 83] and the damping [86, 87] are the most common. Already commercially available devices from the company Q-Sense AB (Sweden) apply the simultaneous measurements of the resonant frequency and of the dissipation of QCRs [87] for cell monitoring applications and the characterization of bio-interfaces through a viscoelastic model. This technique is also known as quartz crystal microbalance with dissipation (QCM-D). Performing impedance spectroscopy on QCRs, however, can provide a wider spectrum of information which can be used to fit more accurate models for the determination of the

1. Introduction

acoustic load applied to the sensor. In fact, former studies showed indeed that QCR impedance spectroscopy is a powerful tool for the analysis of visco-elastic properties at the cell-sensor interface in order to monitor the kinetics of cell attachment [6, 8, 11]. For a single homogeneous layer onto the QCR, the acoustic load $Z_L(\omega)$ can be calculated based on complex shear modulus G_L, density ρ_L and thickness d_L [8, 9, 10]. With the measurements being performed in contact with liquid (water in first approximation) the QCR is operated in the so called non-gravimetric regime. For a 10 MHz resonator the decay length of the shear wave generated in water is of about 180 nm which is considerably less than one cell diameter (~10 µm). Therefore, the shear wave propagation of the QCR is mainly governed by the processes taking place within the region where the actual cell adhesion occurs. Thus, changes in the cellular acoustic load $Z_L(\omega)$ can be mainly attributed to cell motility including alterations of the cell cytoskeleton properties. The online analysis of cellular adhesion kinetics on QCRs in micro channels has recently gained increasing interest [11, 12] and epithelial cells have been cultivated and stimulated inside a micro fluidic bioreactor in flow-through and overpressure regime. Results obtained using a single QCR led to the conclusion that the cellular response may depend on the cell distribution on the sensor surface. That is, even though the experimental conditions and the initial confluency (the surface fraction covered by cells) on the QCR between consecutive experiments are constant, statistical variations in the arrangement of single cells or cluster of cells lead to a different growth of the overall cell layer. Therefore, a biosensor array with multiple QCRs where different cell populations are simultaneously growing under the same environmental conditions was developed. Additionally, the device was realized with disposable parts in order to improve cost and efficiency by decreasing the risk of contamination and time consuming washing procedures. For a comparison, one of the measurement stations commercialised by Q-Sense allows simultaneous data acquisition on four flow modules which can be connected in series or in parallel by the user through capillaries and high-pressure liquid chromatography (HPLC) fittings. Each flow module houses only one resonator and it is plugged on a measurement station which holds all the modules. This solution does not employ any disposable component. Moreover, due to its single-units architecture and the need of external connections between them it may occur that especially at low flow rates the units are not affected by the same environmental conditions. Up to now, there exists no other disposable micro-fluidic biosensor array which can be applied for parallelized infection/stimulation analysis of adherent cell cultures where different cell populations can simultaneously grow under the same environmental conditions within a single unit.

1. Introduction

Electrical impedance spectroscopy using microelectrodes was combined on the same device for the measurement of trans-epithelial electrical impedance across the cell monolayer. This enables collecting of further information for the characterization of the biological system. By measuring the electrical impedance over a large frequency range it is possible to distinguish between changes of the cell monolayer capacitance and conductivity from changes of the capacitance of the semiconductor electrode, which can be attributed to changes of the surface pH [13]. Due to its characteristics of transparency and conductivity, indium-tin oxide (ITO) is a very promising electrode material for biosensor applications combining optical and electrical techniques [14]. Moreover, thanks to their polarizable properties, ITO surfaces are stable under physiological conditions maintaining high sensitivity without insulating oxide layers [15].

1.2. Objectives

The main objective of the present work was the development of a novel disposable microfluidic device for fast online adherent cell monolayer (e.g. living epithelial cells) analysis during cell proliferation and stimulation combining two independent sensor principles and online light microscopy. The new biosensor array consists of 4 bioreactor units and a flow distribution network embedded within the same device. Each bioreactor unit could house four QCRs. Furthermore, the device was designed with a symmetric structure in order to enable the equal partitioning of the cell suspension and obtain the same initial confluency of the cell layer along with identical proliferation conditions during cell cultivation on the surface of each QCR. Quartz crystal resonators were used for the analysis of visco-elastic properties at the cell–sensor interface, while transparent ITO microelectrodes deposited on the top of the bio-chamber, enabled online trans-epithelial electrical impedance measurement in the kHz and MHz range. The latter was used to study dielectric and conductive properties of the cell monolayer as well as the culture medium/stimulus. Electrical impedance spectroscopy technique was employed to determine the equivalent circuit parameters of all QCRs and microelectrodes. A sensor interface electronics including a multiplexer for 32 sensors (16 QCRs and 16 microelectrodes) was developed which enabled fast and parallelized spectra acquisition with a miniaturized impedance analyzer developed in-house [16]. The array was operated in flow-through and overpressure regime. An external flow injection system was used for automated and parallelized media feed as well as the control of the overpressure regime. This approach enables well defined proliferation and stimulation conditions. Besides, it prevents the accumulation of contaminants. Madin-Darby Canine Kidney (MDCK) cells, a permanent non-tumor cell line derived from dog kidney, were chosen for the experimental activity with the biosensor array, because they have a distinctive epithelioid morphology and display several functional and anatomical properties of normal epithelial cells. MDCK cells represent the only widely used model cell line for studies on epithelial polarization, formation and regulation of cell-cell contacts and cell motility [1]. COMSOL simulation results were employed to prove that the shear stress on the adherent cell layer due to the flow-through regime can be neglected and also that the fabrication accuracy has no significant effect on the experimental conditions. This, validated the assumption that the liquid flow is equally partitioned within the symmetric micro-channel network and supported the hypothesis of identical cultivation conditions for the cells living above the sensors

Thin-film deposition techniques were applied for the fabrication of sensors [19, 20]. Injection molding technology [17, 18] was chosen for the cheap mass production of disposable devices. Furthermore, the injection molding process was simulated in order to optimize the mold geometry and minimize the shrinkage and the warpage of the parts. The experimental results suggest that the developed biosensor array could be successfully employed as a support to standard techniques for acquiring deeper insight into that complex system of communication that governs basic cellular activities and coordinates cell actions.

2. CELLULAR FUNCTION AND RESPONSE

The cell membrane represents the most important element of the cell as it relates to this research and it will be the focus of the present chapter. However, understanding of the cell biology cannot be entirely achieved without necessary knowledge of biochemistry, molecular biology, genetics and metabolism. Clearly, a complete review of these subjects is not the object of this work. Therefore, the present chapter will only focus on those aspects of the cellular biology which are related to the physical properties that can influence the sensors responses. The morphology of the eukaryotic cytoskeleton along with the mechanisms of cellular adhesion and motility [4, 21 - 29] will be described in order to provide an intuitive understanding of how the acoustic load on a QCR can be influenced by a layer of adherent cells. Furthermore, the electrical impedance characteristics of the cellular membrane [30 - 34] will be introduced as they represent the main contribution to the signals obtained from the trans-epithelial electrical impedance measurements across the cell monolayer.

2.1. Cellular morphology

A living cell is a very complex system consisting of an external semi permeable membrane which encloses several organelles (nucleus, lysosomes, centrioles, mitoconria, etc.) embedded into an intracellular fluid called cytoplasm as it is shown in figure 1.

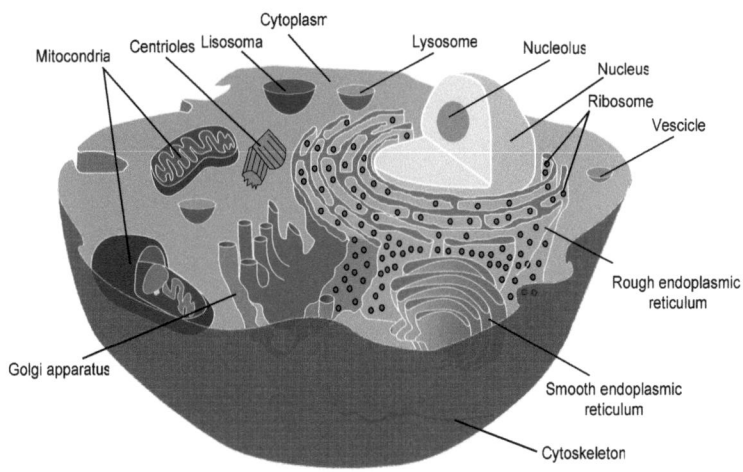

Figure 1 - Schematic of a typical animal (eukaryotic) cell.

2. Cellular function and response

The cell membrane that encloses the cell, has a thickness of about 7 – 10 nm and it mostly consists of a lipid bilayer composed of fatty acid chains. These molecules have a hydrophilic (attracted to water) head and hydrophobic tail (repelled by water) and they spontaneously orient themselves with head groups at both ends and fatty tails facing each other as shown in figure 2.

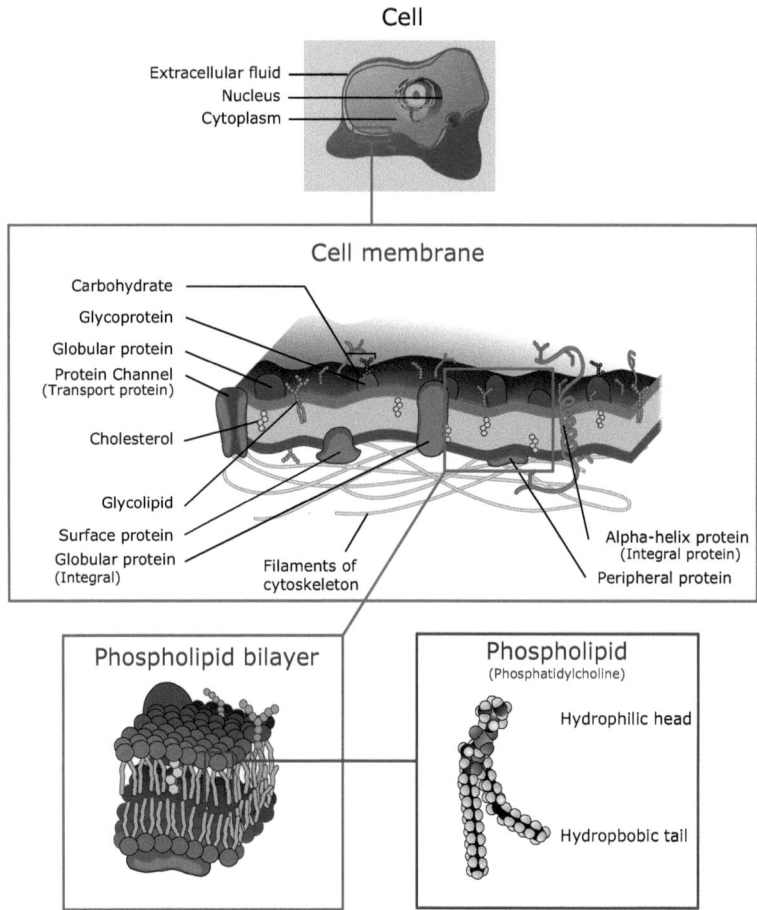

Figure 2 - Cell membrane structure

On both sides, the membrane is in contact with water and its special construction makes possible that the cell can survive and maintain independent intracellular cytoplasm in an aqueous environment. Electric properties of the membrane and bioelectric phenomena derive mainly from the transport of ions through the membrane. The latter is generated and regulated by pore-forming proteins embedded into the membrane which can act as channels and pump for transport of ions and other molecules. Furthermore, included within the membrane there are several other molecules that provide structural integrity to the membrane itself. As a matter of fact, not only proteins but also cholesterol molecules provide some rigidity to the membrane. However, cellular shape along with adhesion and movement are mainly defined by a network of protein filaments (classified as structural proteins) which forms the so called cytoskeleton.

2.2. The eukaryotic Cytoskeleton

The cytoskeleton is a cohesive meshwork of proteins filaments which constitute the cellular "scaffold" and it is responsible for the cellular shape and mechanical stability. It also allows motility and adhesion of the whole cell on substrates as well as intracellular transport and movement inside the cell. Even though the cytoskeleton is built up by a set of various interrelated proteins that operate together, for the purpose of this study, it can be considered consisting of three types of structural protein filaments: actin filaments, microtubules, intermediate filaments as shown in figure 3.

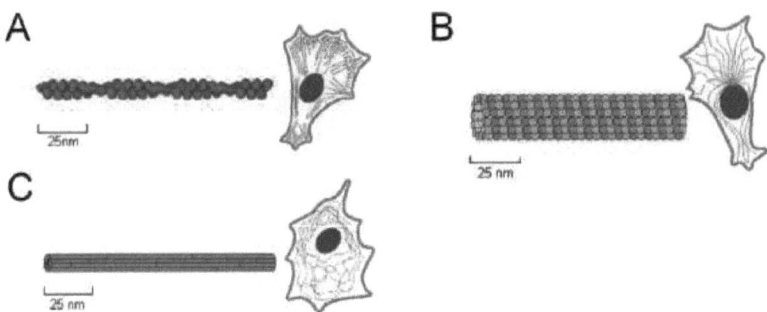

Figure 3 - Principal types of structural protein filaments. (A) Actin filaments. (B) Microtubules. (C) Intermediate filaments [4].

Since they are mostly responsible of the cellular shape, *actin filaments* are mainly present underneath the cellular membrane as actin network that consist of loosely packed interlaced filaments. Actin filaments are two-strained helical polymers of the protein actin and have a diameter of about 7nm. They hold membrane-spanning proteins in their place and they actively participate to cell motility by means of polymerization and depolymerization. Actin filaments also organize in form of bundles which consist of strongly packed arrays of fibers and generally project from the cell creating fingers of membrane. Among these, filopodia are responsible of cell adhesion as they establish contact to the surface beneath the cell.

Microtubules are long empty cylinders of 25 nm diameter formed by polymers of the protein tubulin. Instead of organizing themselves in from of networks, like actin filaments, microtubules form long and straight structures originating from the nucleus and making contact with the actin network. They are responsible of intracellular transport of different organelles and play important role in the organization of cell movements.

Intermediate filaments have intermediate size between actin filaments and microtubules and their main function is of mechanical support as reinforcement of the cell membrane while the cell changes shape during cell movement.

2.3. Cellular Adhesion

The extracellular matrix is a mixture of different proteins and polysaccharides locally secreted by the cell and assembled to form an organized network. It is in the context of this matrix that the bridge between the substrate and the cellular membrane is generated. Such bridge realizes the adhesion of cells to the substrate and connects the cell's cytoskeleton to the extracellular matrix. As a result, cells bind to the culture substrate at precise locations known as focal contacts which held the cell membrane at about 15 nm from the underlying substrate. It is at these sites that the cell cytoskeleton is actively coupled to the extracellular space. More in detail, the adhesion of the cell to the extracellular matrix is realized through integrins, a big family of membrane proteins that mediates cellular contacts to the extracellular matrix. Such proteins are linked, through a chain of other proteins, to actin filaments from the intracellular side, and bound to vitronectin and fibronectin in the extracellular space, as shown in figure 4.

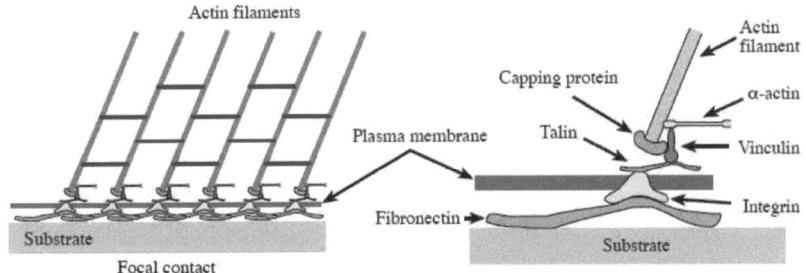

Figure 4 - Structure of focal contacts

Focal contacts in addition, function as transducer of signals from the extracellular matrix to the cell interior, thus they actively contribute to the regulation mechanisms responsible for cell motility, cell polarity, cell growth and survival.

2.4. Cellular Motility

The role of structural proteins in cell structure and morphology has been discussed briefly in section 2.2. In the same frame, it was also pointed out that polymerization and depolymerization of protein filaments are the basis of cell motility; however there was no further explanation about how these phenomena can result in cell movement. As a matter of fact, what cells do while moving is to convert chemical energy into mechanical force and for this reason there are special enzymes responsible for such conversion. By binding and moving along the actin and tubulin filaments, these so called motor proteins generate mechanical forces that make the filaments slide against each other in opposite directions resulting in cellular contraction. Therefore, the motion of cells is the result of polymerization and depolymerization of tubulin and actin filaments in combination with the activity of motor proteins, and the formation of focal contacts. As shown in figure 5, cell motility manifests itself with the occurrence of cell crawling by which cells move or slide on a flat substrate. Cell crawling, can be represented as the cyclic repetition of three different stages: protrusion, attachment and contraction. During protrusion, actin polymerization extends the leading cell margin creating microspikes and lamellipodia, consequently the cell's structure gets stretched and the edge moves forward. Meanwhile, the production of new focal contacts, containing integrin, is carried out and the leading edge of the cell attaches to the substrate. Finally, during contraction, the cell uses the newly generated focal contacts as an anchor to pull (or push)

2. Cellular function and response

itself over the underlying surface. Focal contacts at the trailing edge of the cell are released in order to allow the movement.

It is of course clear that the details of cell attachment and motility are cell specific. For the purpose of this research, it is however assumed that the simplistic model proposed herein can give unambiguous insight into the mechanisms of cell motility and adhesion which give significant contribution during the online analysis of cellular adhesion kinetics on QCRs, namely the analysis of visco-elastic properties at the cell–sensor interface.

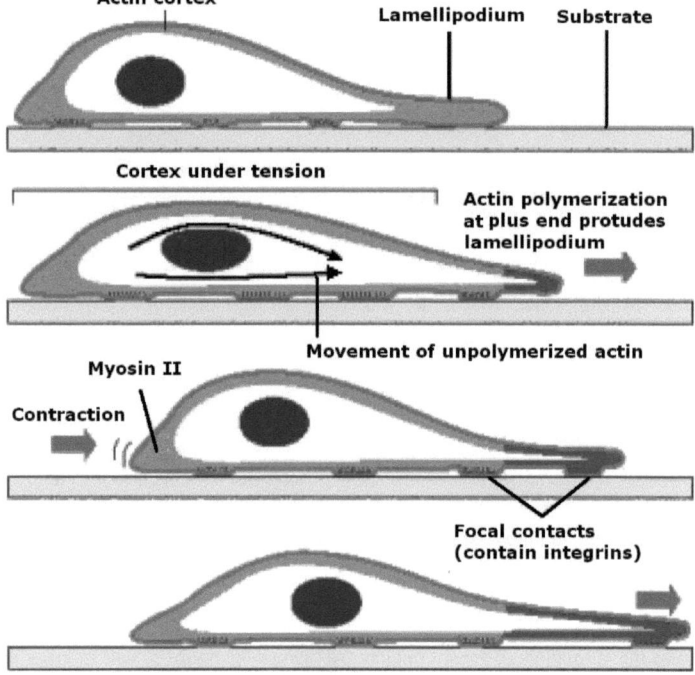

Figure 5 - Steps during cell crawling over a surface [4].

2.5. Membrane Impedance

From an electrical point of view, the cellular membrane can be seen as an insulating layer that separates two conductive solutions. Therefore, its capacitance can be calculated from the area (A) and thickness (d) of the layer along with the relative permittivity of the material (ε_r) using the following relation:

$$C = \frac{\varepsilon_r \varepsilon_0 A}{d} \quad (1)$$

where ε_0 is the permittivity of vacuum (8.85e^{-12} C/Vm) and ε_r is the membrane permittivity. As a result, with a total lipid bilayer thickness of 8 nm, the membrane capacitance has a value of approximately 0.01 pF/µm^2 for most of biological membranes. As well, due to the presence of ionic channels permeating the membrane, there is a resistivity behaviour and the total resistance of the cellular membrane is given by the parallel combination of all the ionic channels which are open at a given time. Moreover, approximating a single ionic channel with a cylinder of known length (l) and cross-sectional area (A) filled with a solution of known electrical resistivity (ρ_e) the resistance R can be estimated as:

$$R = \frac{\rho_e l}{A} \quad (2)$$

The ion channel density is relatively low and the resulting membrane resistance per unit area ranges from 1 MΩµm^2 up to 100 GΩµm^2 depending of the cell type without significantly alter the capacitance. Thus, the simplest model of a cellular membrane is a parallel RC circuit.

However, when actively driving the membrane over a wide frequency range it is important to consider also the dispersive behaviour of biological materials. At low frequencies, they express very high dielectric constants which fall off in fairly distinct steps with increasing excitation frequency. As well, the conductivity increases in steps with increasing frequency. H.P. Schwan [30] showed that the electrical properties of biological materials are characterized by three distinct regions of dispersion named α, β and γ which are placed at low frequency (50 to 200 Hz), radio frequency (0.1 to 10 MHz) and microwave frequency (> 1 GHz) respectively. Different mechanisms of dispersion account for each region [31, 32]. The first dispersion is caused by polarization of the counterion atmosphere surrounding the cell. The β dispersion has been recognized as a Maxwell-Wagner relaxation due to electrical charging of the cell membrane. The last region of dispersion originates from the dispersive behaviour of water molecules which are abundantly present in biological tissues. Finally, each region of dispersion can sometimes be further differentiated and express multiple relaxational behaviours due to additional smaller magnitude fine structure effects.

3. FUNDAMENTALS

The following chapter will introduce the basic concepts and the theory necessary for the understanding of the sensors' principles and the modelling of the sensor data acquired during the experiments. The chapter is divided in three main sections. The first introduces the AT-cut quartz crystal resonators along with the classical models used for the impedance analysis. The second section focuses on microelectrode impedance spectroscopy applied for the measurement of the trans-epithelial impedance of the cell layer during cultivation experiments. Models for the interface between the electrode-electrolyte interface and for the cell layer will be proposed. Finally, the third section will describe the injection molding technology applied for fabrication of cheap disposables.

3.1. The Quartz Crystal Resonator

The Quartz Crystal Resonator (QCR) is a resonant device consisting of a piece of piezoelectric material. It is realized by precisely cutting a slab from a single crystal of quartz (figure 6A) with specific orientation with respect to the crystallographic axes. Due to the piezoelectric effect, internal mechanical stress is produced within the material, and therefore the crystal can be induced to vibrate, when a pair of conducting electrodes is affixed to the QCR unit and periodic voltage is applied. Furthermore, when the QCR is made to oscillate to one of its mechanical resonances, the amplitude of such vibration, will reach a maximum.

As with all solid structures a quartz crystal resonator can express different kinds of bulk acoustic wave modes at the resonance frequencies. Note that the crystal can also oscillate at overtones of every fundamental mode and that the existing modes can sum up to form somewhat complicated resonances modes. Therefore, it is preferably to have the possibility to select only one particular mode and suppress the unwanted ones so that the resonator is oscillating at only one fundamental mode. This selection demands the crystal slab to have proper shape and be cut at a specific crystallographic orientation. The AT-cut quartz crystal, which is used in the present work has a fundamental frequency of 10 MHz, and belongs, together with the BT-cut, to the rotated Y-cut family. Such orientation oscillates in the thickness shear modes and is obtained by rotation of the Y-cut plate about the X-axis as shown in figure 6B.

3. Fundamentals

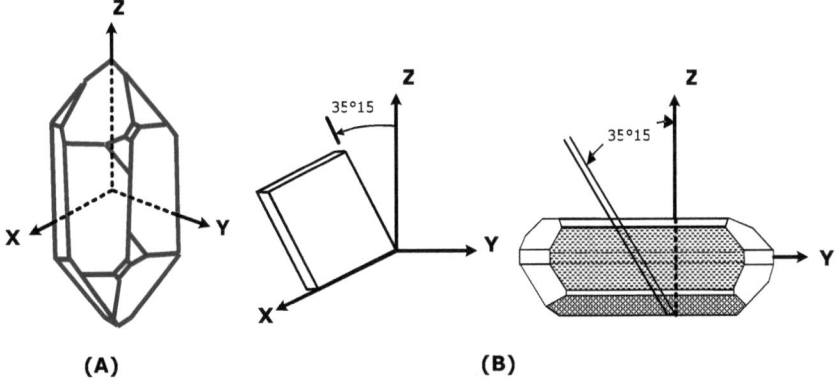

Figure 6 - The quartz crystal. (A) Assignment of axes to the quartz crystal. (B) AT-cut quartz crystal plate.

The AT-cut, also reveals very good temperature stability which is due to the presence of temperature coefficients of the elastic stiffness constants compensating each other at some specific angles of cut. Thus, it is possible to obtain high degree of frequency stability under changing temperatures by precisely controlling the angle of cut. Therefore, the reference angle for the AT-cut slightly varies with design, i.e., with overtone, plate contour, temperature coefficient of frequency etc, and it has a value of about -35°15' to the Z-axis of the crystal.

3.1.1.   Model of the sensor

For a generic acoustic-wave sensor, the basic principle of operation consists of a wave travelling within a confining structure so that a standing wave with distinct frequency is produced. The geometrical dimensions of the confinement structure together with the travelling velocity of the wave, determine the resonant frequency of the device. The QCRs used in this work have a resonant frequency of 10MHz which leads to a thickness value of 166µm. The diameter of the crystal plate is 8mm, whereas the diameters of the ground electrode, faced to the liquid phase, and the working electrode are respectively 6.5mm and 3mm.

3.1.1.1.   The thickness shear mode and acoustic load concept

The geometrical boundaries that concern to the AT-quartz crystal resonator employed in this study consist of a pair of circular electrodes placed on the main surfaces of the disk so that, when a voltage is applied, the resulting electrical field is parallel to the resonator plate thickness. Therefore, the thickness shear mode that arises due to piezoelectric properties and

3. Fundamentals

the crystalline orientation consists of a clean shear deformation of the crystal in X-direction (fig 6). The resulting standing wave is perpendicular to the main surfaces [35, 36], as it is shown in figure 7a. The relation that describes the frequency f_0 of the resulting mechanical resonance in relation with the acoustic velocity v_q of the propagating plane wave and the crystal thickness d_q is the following:

$$f_0 = N\frac{v_q}{2d_q}, \quad N = 1, 3, 5, \ldots \quad (3)$$

In this context, for the purpose of modeling, the shear vibration resulting from the application of alternating voltage across the crystal is treated as an acoustic wave propagating in a multilayer structure. This phenomenon is the physical background of the Acoustic Load Concept (ALC) by means of which an acoustic-electrical behavior including the resonance phenomenon can be derived for the quartz crystal resonator [36 – 41]. The multilayer arrangement is generally composed by a stack of different non-piezoelectric layers joined to a piezoelectric layer (the crystal resonator plate). Thus, in this configuration, the acoustic wave propagates through all adjacent layers, translating and deforming them, thereby probing their acoustical properties. Therefore, the resulting electrical impedance that can be observed at the QCR's electrodes is influenced by the acoustic load due to the superposition of all layers. In the present work, the stack is constituted by the resonator plate, the conducting electrodes deposited on the main surfaces, and the cell culture medium, which is in contact with the big electrode. The acoustic load due to the conducting electrodes can be assumed to be constant, thus observed changes of the electrical response of the QCR can be mainly attributed to variations of acoustic properties of the cell layer adhering to the electrode. Note that the sensitive domain is restricted within the penetration depth δ of the acoustic shear wave into the liquid:

$$\delta = \sqrt{2\eta/(\rho\omega)} \quad (4)$$

Using $\omega = 2\pi*10\text{MHz}$ and the viscosity (η) and density (ρ) of water the penetration depth is found to be about 180 nm. Therefore, the acoustic wave decays before reaching the surface of the liquid and the liquid film can be treated as semi-infinite (figure 7b).

3. Fundamentals

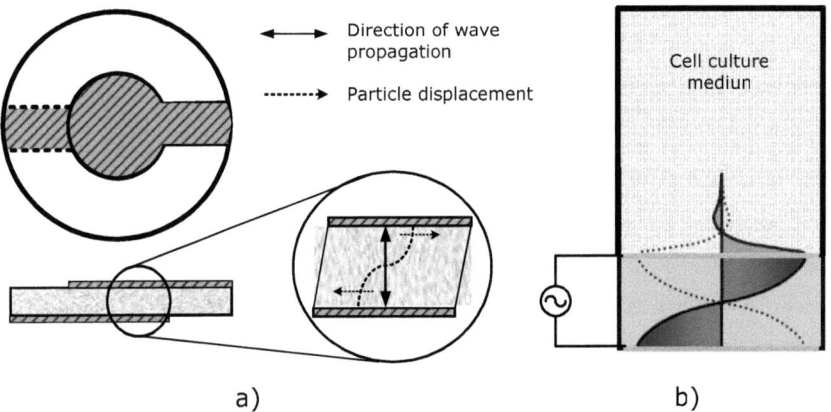

Figure 7 – The acoustic-wave sensor. (a) thickness shear mode. (b) QCR in contact with a liquid.

Finally, for the special case of a QCR presented here, the acoustic load impedance (Z_L) of a single non piezoelectric coating coupled with the resonator is determined by the load density ρ_L and the complex shear modulus G_L and it is defined as:

$$Z_L = j\sqrt{\rho_L G_L} \tan\left(\omega\sqrt{\frac{\rho_L}{G_L}}d_L\right) \tag{5}$$

Penetration of the acoustic wave into the coating can therefore be employed to probe its mechanical properties along with its thickness and the acoustic properties at the upper film surface. The information about energy stored into the medium due to the elastic behavior is found in the imaginary part of Z_L, whereas the part relative to kinetic energy, which cannot be reversibly transformed (dissipated), is represented by the real part of Z_L and refers to the attenuation of the acoustic shear wave.

3.1.1.2. The Transmission-Line Model

The propagation of acoustic waves in the multilayer structure introduced requires applying the theory of wave propagation and it is usually presented as transmission line in analogy to electrical waves. Supposing all layers homogeneous and considering the high aspect ratio between the quartz disc and its thickness, it is reasonable to derive the wave propagation in a one-dimensional model [38, 40]. Therefore, a primary correlation of the acoustic wave propagation and the electrical impedance, measured at the electrodes of the sensor, can be formulated by use of the so called Transmission-Line Model (TLM). Using this equivalent circuit approach, the wave propagation is described in analogy to electrical waves travelling across the device thickness and into the adjacent layers. Moreover, as it will be explained, near resonance it is possible to develop the model into a particular notation, where physical parameters can be summarized in lumped equivalent electrical values.

By means of the TLM, the crystal resonator is represented as a three port black box. Two of them (E-F, G-H) are acoustic ports and represent the main surfaces of the crystal plate. The third one (A-B), is the electrical port which corresponds to the contact electrodes. The transformation of mechanical displacement into the electrical signal and vice versa, or in other words the coupling between the acoustic variables and the electrical port, is performed through the transformer. As well, every non-piezoelectric adjacent layer is represented as a two-port transmission line and a generic load can be attached at the end of the line. Finally, the resulting acoustic load Z_L which is seen at port E-F of the transmission line (shown in figure 8) would be the result from the contribution of all layers placed on the quartz surface. Thus, the acoustic load summarizes all acoustically relevant information and it does not play any role whether this load is generated by a single coating, by a multilayer arrangement or a semi-infinite material. For this reasons, any change in the acoustic load impedance would result in a change of the electrical impedance seen at the electrical port of the QCR. Therefore, these changes are responsible for the frequency shift and attenuation change of the acoustic device. Note that the electrodes are acoustically thin layers and their contribution can be considered constant during the experiments, therefore their effect is included into the quartz block by introducing an effective quartz crystal thickness. Figure 8 represents the common situation where an acoustic load is acting at one acoustic port (E-F) while a stress-free surface is connected at the acoustic port G-H.

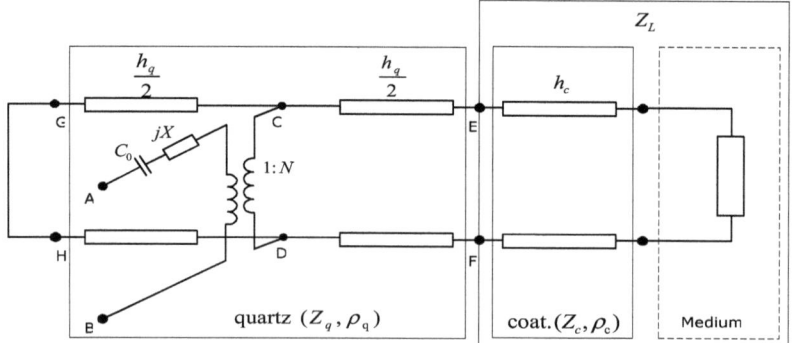

Figure 8 – Schematic representation of the transmission line model

The electrical response of such configuration can be expressed as [41]:

$$Z_{el} = \frac{1}{j\omega C_0}\left(1 - \frac{k_q^2}{\alpha_q} \frac{2\tan\left(\frac{\alpha_q}{2}\right) - j\frac{Z_L}{Z_{cq}}}{1 - j\frac{Z_L}{Z_{cq}}\cot(\alpha_q)}\right) \quad (6)$$

where

$$Z_{cq} = \sqrt{\rho_q G_q}$$
$$\alpha_q = \omega d_q \sqrt{\frac{\rho_q}{G_q}} \quad (7)$$

are respectively the characteristic acoustic impedance of the quartz and the phase shift of the wave within the crystal. G_q is the quartz complex shear modulus and it is in general a complex value:

$$G_q = G_q' + jG_q'' \quad (8)$$

where G_q' is the so-called shear storage modulus accounting for acoustic energy storage and G_q'' is the shear loss modulus accounting for acoustic energy dissipation. C_0 represents the static capacitance created by the parallel-plate capacitor due to the electrodes placed on the main surfaces of the crystal at a distance d_q. Considering the dielectric permittivity of the crystal ε_{22} and an effective area A_{EL} for the electrodes:

$$C_0 = \varepsilon_{22} \frac{A_{EL}}{d_q} \quad (9)$$

Finally, k_q^2 is the electromechanical coupling coefficient which can be calculated as 8.8% for the thickness shear mode in the AT-cut crystal [40, 42, 43].

3.1.1.3. *The modified Butterworth van Dyke Model*

The expression (4) representing the electrical response seen at the electrical port of the QCR can be rewritten in order to describe the impedance Z_{el} as the result of a parallel circuit consisting of a static capacitance C_0 and a motional impedance Z_m. Moreover, the motional impedance can be further split into two parts Z_m^0 and Z_m^L representing respectively the

3. Fundamentals

unloaded quartz ($Z_L = 0$) and the load applied. As well, near resonance it is possible to approximate the acoustic phase shift with $\alpha_q \approx N\pi$ so that:

$$\tan\left(\frac{\alpha_q}{2}\right) \approx \frac{4\alpha_q}{N^2\pi^2 - \alpha_q^2} \tag{10}$$

Consequently, the approximated electrical quartz impedance Z_{el} can be represented, in the vicinity of the resonance, by means of equivalent circuit elements in the so called modified Butterworth-Van Dyke Model, shown in figure 9. The only real electrical value is the static capacitance C_0(7) whereas the capacitance C_{ext} is due to the parasitic introduced by the electrical contacts along with connection to measurement instruments and surrounding medium.

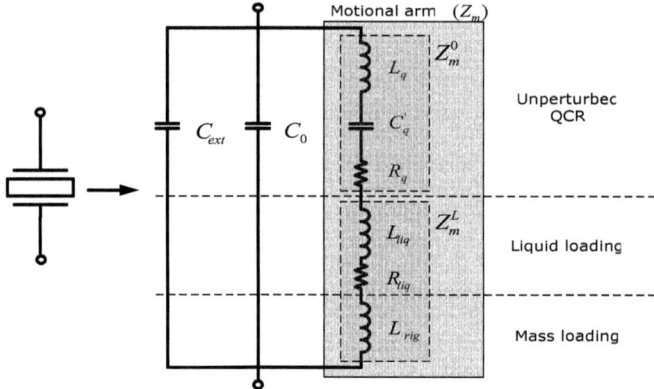

Figure 9 – The modified Butterworth-Van Dyke Model with simultaneous mass and liquid loading [9]

Finally, the lumped equivalent electrical values of the motional arm, for the unperturbed QCR motional impedance Z_m^0, depending on geometric and acoustic sensor properties, are defined as follow [40, 44]:

3. Fundamentals

$$C_q = C_0 \frac{8k_q^2}{N^2\pi^2} \tag{11}$$

$$C_q' = \frac{C_q}{1 - \frac{8k_q^2}{N\pi^2}} \tag{12}$$

$$L_q = \frac{\pi^2}{8C_0 k_q^2 \omega_0^2} = \frac{\rho_q d_q^3}{8A_{EL} e_q^2} \tag{13}$$

$$R_q = \frac{\eta_q}{c_{q0}} \frac{1}{C_q} = \frac{\eta_q \pi^2}{8C_0 k_q^2 c_{q0}} \tag{14}$$

3.1.1.4. The load motional impedance

The imaginary and real parts of the acoustic load impedance are directly related respectively to the frequency shift Δf of the resonant frequency and to changes ΔR of the resistance in the motional arm of the BDV equivalent circuit:

$$\Delta f = -f_0 \frac{\text{Im}(Z_L)}{\pi Z_{cq}} \tag{15}$$

$$\Delta R = 2\omega_0 L_q \frac{\text{Re}(Z_L)}{\pi Z_{cq}} \tag{16}$$

In terms of Eq. (5), when a <u>thin and rigid film</u> is deposited on the QCR it reflects into a real and large G_L and a small d_L. In this situation, the device works in the so called gravimetric regime. Furthermore, considering that the phase shift α within the layer is small enough to approximate the tan-function by its argument the acoustic load of a thin rigid layer is:

$$Z_{rig} = j\omega \rho_L d_L \tag{17}$$

where the only parameter influencing the load is the areal mass density of the layer $\rho_L d_L$ and the electrical response it is only influenced by a mass change on the surface. Applying (15) and (16) leads to an equation very similar to what shown in 1959 by Sauerbery [45]:

$$\Delta f_{rig} = -2f_0^2 \frac{\rho_L d_L}{Z_{cq}} \tag{18}$$

3. Fundamentals

$$\Delta R_{rig} = 0 \tag{19}$$

which means (in principle) that a rigid coating does not add acoustic losses. Finally, even though in the general case (5) cannot be represented by means of single lumped elements of the BDV-model, the contribution to the load motional impedance due to rigid loading can be described with the inductance L_{rig} in series with L_q:

$$L_{rig} = \frac{\pi \rho_L d_L}{4 C_0 k_q^2 Z_{cq} \omega_s} \tag{20}$$

A second kind of typical load commonly applied to the QCR consists of a <u>semi-infinite Newtonian liquid</u> which only attenuates the acoustic shear wave without showing any elastic behavior. This situation is also known as non-gravimetric regime. In this case, G_L is only characterized by the shear loss modulus so that $G_L = G_L^{''} = j\omega \eta_{liq}$. Moreover, due to the semi-infinite assumption $d_{liq} \gg \sqrt{2\eta_{liq}/(\rho_{liq}\omega)}$ and the tan-function in (5) equals to one. Therefore, the surface acoustic impedance generated by a pure viscous liquid loading only depends, by a linear relationship, on the density-viscosity product:

$$Z_{liq} = \sqrt{j\omega \rho_{liq} \eta_{liq}} = (1+j)\sqrt{\frac{\omega \rho_{liq} \eta_{liq}}{2}} \tag{21}$$

Here Z_{liq} has both nonzero real and imaginary part that is to say that liquid loading expresses both frequency shift and increased damping. Applying again (15) and (16) leads to [46 – 48]:

$$\Delta f_{liq} = -f_0^{3/2} \frac{\sqrt{\rho_{liq} \eta_{liq}}}{\sqrt{\pi} Z_{cq}} \tag{22}$$

$$\Delta R_{liq} = 4 f_0^{3/2} L_q \frac{\sqrt{\pi \rho_{liq} \eta_{liq}}}{Z_{cq}} \tag{23}$$

Finally, also in this case Z_{liq} can be represented by means of single lumped elements in the motional arm of the modified BDV-model [9]:

$$L_{liq} = \frac{\pi}{4C_0 k_q^2 Z_{cq} \omega_s} \sqrt{\frac{\rho_{liq} \eta_{liq}}{2\omega_s}} \qquad (24)$$

$$R_{liq} = \frac{\pi}{4C_0 k_q^2 Z_{cq}} \sqrt{\frac{\rho_{liq} \eta_{liq}}{2\omega_s}} \qquad (25)$$

The last case considered here is when a QCR with a thin rigid layer works in a Newtonian liquid. In these circumstances the acoustic loads are approximately additive and the total acoustic load can be obtained from superposition of the two single acoustic loads:

$$Z_{rig+liq} = Z_{rig} + Z_{liq} = j\omega\rho_L d_L + (1+j)\sqrt{\frac{\omega \rho_{liq} \eta_{liq}}{2}} \qquad (26)$$

Also frequency shift and increased damping can be derived in the same fashion, from superposition of single contributions due to thin rigid film and semi-infinite Newtonian liquid.

3.1.2. Impedance analysis of the shear wave resonator

The acoustic load impedance Z_L, and consequently also Z_{el}, is directly related with the frequency shift and the change in the motional resistance. When only Z_m^0 is present in the motional arm, the resonance frequency can be easily calculated as

$$f_{res,0} = \frac{1}{2\pi\sqrt{L_q C_q}} \qquad (27)$$

And the motional conductance maximum is

$$G_{max,0} = G_{el}(f_{res,0}) = \text{Re}(Z_{el}(f_{res,0}))^{-1} = R_q^{-1} \qquad (28)$$

When an acoustic load is present the load motional impedance $Z_m^L = R_{liq} + j\omega(L_{rig} + L_{liq})$ comes into play in the motional arm of the equivalent circuit and the resonance frequency is decreased to

$$f_{res} = \frac{1}{2\pi\sqrt{(L_q + L_{rig} + L_{liq})C_q}} \qquad (29)$$

while the conductance maximum becomes

$$G_{\max} = G_{el}(f_{res}) = (R_q + R_{liq})^{-1} \tag{30}$$

Therefore, characterization of the acoustic load applied to the QCR can be performed by following changes of the resonance frequency and conductance maximum acquired through impedance analysis of the crystal resonator [9, 10].

3.2. Microelectrode impedance spectroscopy

Impedance spectroscopy (IS) is a method of characterizing many of the electrical properties of materials and their interfaces with electronically conducting electrodes. It may be used to investigate the dynamics of bound or mobile charge in the bulk or interfacial regions of any kind of solid or liquid material. Electrical measurements to evaluate the electrochemical behavior of electrode and/or electrolyte materials are usually made with cells having two electrodes applied to the faces of a sample. The general approach is to apply an electrical stimulus to the electrodes and observe the response [50].

Figure 10 shows a schematic draw of the trans-epithelial impedance measurement applied in this work. The ITO electrodes are deposited on the glass cover placed on top of the biosensor array. The voltage applied at the ITO microelectrode generates the electric field which crosses the measurement chamber through the culture media and reaches the cell layer lying on the QCR's ground electrode underneath.

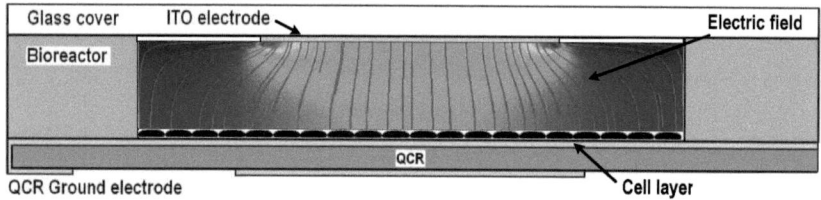

Figure 10 – Schematic draw of the trans-epithelial impedance measurement with simulated electric field

The trans-epithelial impedance spectroscopy of the cell layer proliferating within the biosensor array acts as a function of several parameters and the correlation between the impedance and the surface coverage is quite complicated. However, the cell layer, as well as the metal-electrolyte interface can be modelled using passive circuit elements. Therefore, changes of the measured electrical impedance over a well defined frequency range can be related to the parameters changes of the equivalent circuit, which reflect the status of the cell-electrode interface, i.e. the cell-electrode gap change, as well as the electrochemical equilibrium at the metal-electrolyte interface.

3.2.1. The metal-electrolyte interface

A typical circuit used in electrochemistry [50] to describe the metal-electrode interface is shown in Figure 11.

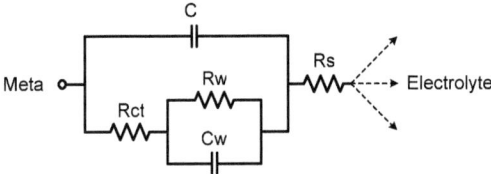

Figure 11 – Equivalent circuit commonly used in electrochemistry

The capacitor C_I represents the interfacial capacitance accounting for electrical double layer and the diffuse layer arising at the metal/electrolyte interface. R_{ct} represents the resistance to the charge flow across the interface. Its value may result from the combination of different contributions such as diffusion of reactants to and from the electrode or chemical reactions at the electrode. However, for voltage values less than 50mV peak, the charge transfer through the double layer due to the applied potential tends to dominate the overall current [22]. R_W and C_W are the components of the Warburg impedance which takes into account the movement of ions in response to an applied electric field. The resistor R_s stands for the resistance of the solution through which all the current must pass to reach the ground electrode.

In contrast to C_I and R_{ct} which could be considered as ideal circuits elements, R_W and C_W are not ideal because they change with the frequency. That is, the diffusional impedance tends to zero with increasing excitation frequency because it is more and more difficult for the ions to follow the applied field and the effects of the diffuse ion cloud become less significant. For the materials and the frequency range employed in this work, it was experimentally determined that the contribution of the Warburg impedance could be neglected.

3.2.1.1. Interface capacitance

The model which first described the electrical double layer was developed by Helmholtz in the 1850's. The double layer capacitance C_H described by this model is the consequence of electron transfer reactions which result in the formation of an electrified interface at the boundary between metal and electrolyte. This results in two layers of charge (the double layer) and a potential drop that is confined within the electrolyte to a distance from the electrode known as outer Helmholtz Plane (OHP). The result is equivalent to a simple capacitor with a value determined by the permittivity of the electrolyte ($\varepsilon_0\varepsilon_r$), the area of the interface (A) and the distance of the OHP from the metal electrode (d_{OHP}):

$$C_H = \frac{\varepsilon_r \varepsilon_0}{d_{OHP}} \qquad (31)$$

where C_H is the capacitance per unit area (F/m^2), ε_r is the relative permittivity of the electrolyte and ε_0 is the relative permittivity of vacuum. Assuming, for physiological saline at 25°C, a value of d_{OHP} equal to 0.5 nm and ε_r ranging from 6 to 78 the Helmholz capacitance can be roughly estimated with a value of few pF/µm^2 [51]. This model, however, neglects the observed dependency of capacitance on the applied potential. Therefore, the Helmholz model was modified in 1910 to 1913 by Gouy and Chapman in order to account for this dependency. The new model considers mobile ions at the electrode surface allowing the capacity to change in response to an applied potential. By this assumption and for voltage values less than 100mV peak the voltage drop across the space charge region can be estimated as:

$$V(x) = V_0 e^{\left(-\frac{x}{L_D}\right)} \qquad (32)$$

where V_0 is the potential at the electrode, x is the distance from the electrode and L_D is the Debye length which can be calculated in an electrolyte as follow:

$$L_D = \sqrt{\frac{\varepsilon_r \varepsilon_0 kT}{2 N_A q^2 I}} \qquad (33)$$

where k is the Boltzmann constant, T is the absolute temperature in Kelvin, N_A is the Avogadro number, q is the elementary charge and I is the ionic strength of the electrolyte in mole/m^3. The capacitance per unit area resulting from the Gouy-Chapman model can be calculated by:

$$C_G = \frac{\varepsilon_r \varepsilon_0}{L_D} \cosh\left(\frac{zV_0}{2V_t}\right) \qquad (34)$$

where z is the valence of the ions and V_t is the thermal voltage. The first term of (34) expresses the capacitance per unit area of a parallel plate capacitor and the hyperbolic cosine compensates for the effects of mobile charges.

The total interface capacitance C_I is given by the Gouy-Chapman-Stern model as follow:

$$\frac{1}{C_I} = \frac{1}{C_H} + \frac{1}{C_G} \qquad (35)$$

A detailed theoretical derivation of the interfacial capacitance can be found in [52].

3.2.1.2. Solution resistance

The solution resistance R_S is the resistance measured between the ITO working electrode and the QCR's ground electrode. It can be determined from the spreading resistance which models the effects of the current spreading out into solution from the localized electrode to a distant counter electrode. Its value can be obtained by integration of the series resistance of shells of solution centered at the electrode with the solution resistance given by:

$$R = \frac{\rho_{el} l}{A} \qquad (36)$$

where ρ_{el} is the electrical resistivity of the electrolyte, A is the cross-section area through which the current must pass and l is the length. The spreading resistance of a circular electrode with radius r is given by:

$$R_S = \frac{\rho}{4r} = \frac{\rho\sqrt{\pi}}{4\sqrt{A_{EL}}} \qquad (37)$$

were A_{EL} is the area of the circular electrode [53]

Finally, the complete circuit model assumed for the metal-electrolyte interface is shown in figure 12. This model simplifies the one presented in figure 11 by neglecting the Warburg contribution and it is well suited for electrodes systems in pure electrolytes. However, when proteins or other additional elements are present in solution it may be necessary to use empirical methods to determine the electrode model parameters. Also, inhomogeneous

3. Fundamentals

electrical fields could be generated by the electrode geometry and a more complicated behavior than predicted by the model may be observed in practice.

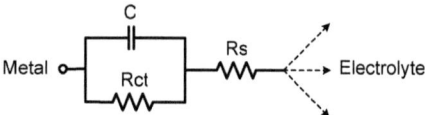

Figure 12 – Circuit model employed for explaining the contribution of the metal-electrolyte interface.

3.2.2. Cell layer electric model

A brief description of the membrane impedance was already given in paragraph 1.5 where a parallel RC circuit was suggested as the simplest way to model the cell membrane in terms of electrical parameters [21, 54]. Typical values for the cell membrane capacitance and resistance were also given. However, for the purpose of modelling the contribution of the cell layer on the impedance change measured at the electrodes, it should be pointed out that the single cell membrane resistance as well as the single cell membrane capacitance cannot be employed directly. The reason is to find in the frequency range at which the impedance measurements are here performed. The impedance spectra are acquired between 100 kHz and 10 MHz in which the β dispersion normally takes place. Therefore, the measured capacitance and resistance introduced by the cell layer show a value that is frequency dependent. As a rule, decreasing dielectric constant and increasing conductivity are responsible for decreasing capacitance and decreasing resistance values respectively.

3.2.2.1. *Relaxation and dispersion*

Assuming the β dispersion being originated from a single relaxation mechanism, this can be described by first order differential equations which lead to single time constant responses [32]. Therefore, when a voltage step is applied to a sample, its polarisation can be characterised as a first order process with a relaxation time τ which depends on the physical process involved:

$$D(t) = D_\infty + \frac{D_0 - D_\infty}{1 - e^{-t/\tau}} \qquad (38)$$

where D_∞ represents the "instantaneous" (high frequency) polarisation of the sample, while D_0 is the response obtained long after the application of the step (low frequency). By Laplace transformation the response in the frequency domain can be obtained. Also considering the real part of the complex permittivity $\varepsilon' = D/E$ and $C = \varepsilon' A/d$ it is possible to show that:

$$\bar{\varepsilon}(\omega) = \varepsilon'_\infty + \frac{\varepsilon'_S - \varepsilon'_\infty}{1 + j\omega\tau} \tag{39}$$

$$\bar{C}(\omega) = C_\infty + \frac{C_S - C_\infty}{1 + j\omega\tau} \tag{40}$$

where the subscript s stands for "static" since ε_0 refers to the permittivity of vacuum and C_0 to the static capacitance of the QCR. According to equations (39) and (40), the permittivity (and consequently the capacitance) drops off in a distinct step going from one constant value ε_S to another ε_∞ over about one decade as the frequency increases. The frequency f_c, where the average value between the two levels occurs, is known as the characteristic frequency. The relaxation time τ can be easily obtained from the characteristic frequency as follow:

$$\tau = \frac{1}{2\pi f_c} \tag{41}$$

An increase in conductivity σ is also associated to the drop in permittivity by the following relation:

$$\sigma_\infty - \sigma_S = \frac{\varepsilon_\infty - \varepsilon_S}{\tau} \tag{42}$$

Due to the distribution of cell sizes and their complex organization in real biological materials a number of relaxation processes often occur in parallel so that the total electrical response could be characterised by more than one time constant. Moreover, if the relaxation times are not enough separated the material will exhibit a single broader dispersion with smoother changes between permittivity values.

3.2.2.2. Maxwell-Wagner effects

The Maxwell-Wagner effect caused by cell membranes has been recognised as the physical basis for the β dispersion. The Maxwell-Wagner effect consists of a relaxation process which takes place when an electric current has to cross an interface between two different dielectrics. In order to model biological systems, the general Maxwell-Wagner theory can be extended to a dilute suspension of spheres surrounded by a thin membrane [32]. However, in order to develop an equivalent electrical model it is assumed that the electrical properties of a cell

3. Fundamentals

layer adhering on the sensor electrode can be better described by a capacitor with two dielectric slabs placed in series between the plates (rather than coated spheres in dilute suspension). Each slab has a resistance R_i in parallel with a capacitance C_i, and a time constant $\tau_i =. R_i C_i$. From the total admittance of such a capacitor and provided that $Y = j\omega C + 1/R$, the resultant capacitance and conductance are:

$$C = \frac{\text{Im}(Y)}{\omega} = \frac{1}{R_1 + R_2} \frac{\tau_1 + \tau_2 - \tau + \omega^2 \tau_1 \tau_2 \tau}{1 + \omega^2 \tau^2} \tag{43}$$

$$\frac{1}{R} = \text{Re}(Y) = \frac{1}{R_1 + R_2} \frac{1 + \omega^2 [\tau(\tau_1 + \tau_2) - \tau_1 \tau_2]}{1 + \omega^2 \tau^2} \tag{44}$$

where:

$$\tau = \frac{R_1 \tau_2 + R_2 \tau_1}{R_1 + R_2} = \frac{R_1 R_2 (C_1 + C_2)}{R_1 + R_2} \tag{45}$$

is the time constant of the whole system and is equivalent to the relaxation time introduced with Eq. 41. The system capacitance and resistance values monotonically decrease with increasing frequency and converge both at low and high frequencies:

$$C_S = \frac{C_1 R_1^2 + C_2 R_2^2}{(R_1 + R_2)^2} \text{ and } R_0 = R_1 + R_2 \quad f \to 0 \tag{46}$$

$$C_\infty = \frac{C_1 C_2}{C_1 + C_2} \text{ and } R_\infty = \frac{R_1 R_2 (C_1 + C_2)^2}{R_1 C_1^2 + R_2 C_2^2} \quad f \to \infty \tag{47}$$

Thus, with the parallel model of two slabs in series it is possible to describe a classical dispersion. The dispersion is due to the parallel of a capacitance with a conductance for each dielectric so that the interface can be charged by the conductivity. This model is shown in figure 13 and will be used for describing the change of the trans-epithelial impedance observed over time during cell proliferation.

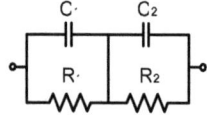

Figure 13 – Circuit model employed for explaining the contribution of the cell layer.

3.3. Injection molding technology

Injection molding is a very cost-effective means for the production of three-dimensional plastic parts with high throughput. Most of the time the produced parts require little or no finishing and very complex shapes can be formed. During the injection molding process solid granules of thermoplastic resins are melted, injected into a hollow cavity space (mold) built to the shape of the desired product, and then cooled back to a solid state in the new desired form. In order to fill the empty form, the mold is mounted in an injection molding machine that is automatically looping through the steps of the molding cycle:

- Closing of the cavity. The clamping mechanism rapidly opens and closes the mold with the necessary clamping force.
- Injection of the molten polymer into the cavity spaces. The injection unit presses the melt into the mold with a pressure that is largely dependent on the conformation of the product and the material being injected. Also, the clamping mechanism must supply sufficient clamping force in order to counteract the pressure rising within the cavity and keep the mold close during this phase.
- Mold pack and hold. During this step, a final fill pressure is maintained in order to pack the cavity and preserve the internal mold pressure until solidification of the part.
- Cooling of the part. Time during which the part is cooled down before ejection. Uniform and properly controlled cooling of the part can help preventing warpage and achieve minimum cycle times.
- Opening of the cavity and ejection of the part. The part should be removed quickly and without damage which could occur if the part is ejected too cold or too hot. Normally, all molds have an automatic ejection mechanism.

3.3.1. Mold design and fabrication

When designing a mold cavity the manufacturing process must be always considered. During the stages of the injection molding cycle there are many factors affecting the quality of the parts as well as the repeatability of the process. Therefore, during part design it is important to bear in mind few important guidelines that will help minimize problems during molding and improve the final result:

- The wall thickness should be minimised as much as possible in order to reduce the effect of shrinkage.

- Large variations and abrupt changes in wall thickness should be avoided for reduction of differential shrinkage which could lead to part deformation and warpage.
- Sharp corners should be avoided for minimisation of stress within the part.
- Parts should be designed with a minimum of 0.5° draft angle in order to facilitate their ejection from the form.

Once the shape of the parts has been defined, additional effort should be given in order to complete the design of the mold cavity. At first, the running system should be included for transporting the melt from the nozzle to the cavity in the most uniform way and with minimum pressure and temperature drops. The gate is the channel through which the molten polymer flows into the cavity. It connects the runner to the part and it is normally a highly stressed area. Therefore, it is of importance to find a gate with geometry and location that do not adversely affect the appearance and properties of the product. A properly designed gate should address the following issues:

- It should ensure a balanced melt flow for rapid and uniform filling of the cavity in order to avoid over-packing of some areas of the part.
- Whenever possible, it should be placed in the thickest section of the part so that the injected material would flow from thick to thin sections.
- It should not be placed near an area with a large variation in wall thickness.
- It should ensure filling of the mold under realistic temperatures and pressures.
- It should minimize weld lines and position them in non-critical areas of the part.
- It should prevent "jetting" by providing smooth and uniform material flow.
- It should avoid air entrapment in the part.
- It should avoid visual distraction from the overall part appearance.

A second issue to be considered when designing the mold cavity is the evacuation of air as the form is being filled during injection. Therefore, vent grooves should be appropriately placed around the part and at the end of the flow paths. Inappropriate venting may be responsible for slow and incomplete fill along with air bubbles and weak weld lines. Moreover, compressed trapped air could heat up enough to ignite and leave burn marks.

As a final point, ejection of the part should be addressed in order to reliably and quickly remove the part from the cavity. Thus, ejection pins must be properly placed in order to actively push the part out of the form. Moreover, it is very important to ensure that as the mold opens, the parts will stay on the side from which they will be ejected. For this reason, if the product will not naturally stay on that side, features such as undercuts should be introduced to ensure that the part will be where required for ejection. In order to ensure good and reliable ejection, the number of pins should be enough to distribute the necessary force so that the part is not affected or damaged. Furthermore, the pins should be as large as possible and evenly placed at symmetrical locations. Also, some of them should be placed near the corners.

3.3.2. Influence of processing parameters

During the injection molding cycle there are many variables affecting the quality of the fabricated product. Only the most significant process parameters will be considered in this frame.

Injection fill rate. It has a big effect on melt viscosity and orientation and consequently on warpage due to differential shrinkage. Because of the augmented shear heating, an increase of fill rate is comparable to raising the melt temperature and it can facilitate the filling of cavities owing to decreased material viscosity. However, a large increase of the shear rates may also result in degradation of the polymer, cavity gassing and hence air bubbles in the product. A good balance between injection speed and shear rate should be found for optimum filling.

Melt temperature. It has direct influence on the polymer flow. With most polymers, an increase of melt temperature leads to a reduction of material viscosity and shear stress along with improved flow into the cavity. Moreover, each polymer has a recommended temperature range and it is important to select an appropriate temperature window for preventing the occurrence of material degradation. Also, the melt temperature has a direct influence on part shrinkage so that decreasing its value the volume shrinkage is reduced.

Mold temperature. Higher mold temperatures can reduce the melt flow resistance. Also, the selection of the mold temperature is important for preventing the premature solidification of the melt and to ensure the complete filling of the cavity.

Packing pressure. How well a part is packed out is of primary importance when considering warpage, shrinkage, and defects such as sink marks. To produce a uniform volumetric shrinkage, the pressure in the cavity must be properly controlled by selecting the optimal packing profile.

4. DESIGN OF DEVICES AND MOLD FABRICATION

The design of such a complex system for online monitoring of dielectric and conductive properties of the cell monolayer along with visco-elastic properties at the cell–sensor interface has to be carried out with several criteria in mind. All features must be designed in view of the fabrication technology that is available, in conjunction with several other issues. Among these, biocompatibility, maintenance of sterility and physiochemical environment (pH, temperature, etc.) during experiments, minimizations of death volume for supply of media and stimuli, packaging, handling, connections and electrical interfaces with low parasitic for extraction of the electrical signals are the most significant. Moreover, all these aspects must be collectively considered and a trade off has to be attained for the best overall solution.

4.1. Resin selection

Defining the polymer to be used for the production of the disposable device is a very important choice which should be done in the early phase of the design process. The selection of the resin must combine application related demands with the influence of the material on the molding process itself. A medical grade of transparent polypropylene from LyondellBasell Polymers was selected for the fabrication of the disposable devices.

Polypropylene (PP) is the resin of choice for many injection molded applications because of its low cost and light weight. However, medical applications like this go far beyond these two characteristics. One key property which makes polypropylene so attractive for the market of medical disposables is that it can be steam or irradiation sterilized and still maintains sufficient physical properties to perform its intended function. Also, it has very low water absorption, it is very resistant to chemicals and it does not easily deteriorate during storage. As an example, most disposable syringes today are made of polypropylene. The drawbacks of polypropylene are that it exhibits a relatively high shrinkage and as a consequence it has a greater tendency to warp and form sink marks. However, these tendencies can be effectively counteracted with a proper design of the part and by adjusting the molding conditions.

4.2. Microfluidic network design and simulation

Identical proliferation and stimulation conditions among all bioreactor units, along with equal flow distribution above all QCRs were the major constraints to be considered for the design. These conditions were of great importance for the assumption that the cell communities, namely the cells proliferating on the surface of a single QCR, in the biosensor array are affected by the same environmental conditions. Moreover, equal flow distribution, was very important in order to obtain the same partitioning of the cell suspension on the surface of each sensor during the cell seeding procedure and provide the same starting condition on each sensor. Therefore, the microfluidic network had to be designed with a symmetric structure in order to provide the same fluidic resistance through parallel microfluidic paths. Also, a compromise between minimum dead volume and minimum shear stress on cells had to be found for the design of the micro-channel network. This means that the maximum allowable shear stress during seeding of cells limits the miniaturization of micro channels. Moreover, the internal volume of the bioreactor unit cannot be too small in order to provide a buffer capacity that can maintain constant environmental condition at the cell layer. Finally, constraints introduced by the injection molding process for the fabrication, e.g. avoiding hidden structures, had to be taken into account. The bioreactor unit with 4 sensors in parallel was developed based on an existing micro fluidic bioreactor for a single QCR ($\varnothing$ 14 mm) [11]. Figure 14 shows how the four bioreactor units and the flow distribution network have been merged within the same device: the flow distribution network lays on the bottom side of the device, whereas the 4 bioreactor units are placed on the top side of the bioreactor array. Each bioreactor unit has two inlets and one outlet and the connections between the two sides are realized through holes. The total internal volume of the device, including the flow distribution network and the 4 bioreactor units, is ~1 ml. Fluid flow is in parallel to the sensor surface to minimize mechanical force on cells. QCRs are fixed to the bottom side of the device using paraffin. The contact area between cells and sensor has a diameter of 4.5 mm. Glass plates cover the bioreactor units and enable the acquisition of light microscopy pictures. The areas highlighted with darker colour on the top side of the device have a thicker cross section compared to the rest and allow the correct alignment of the glass plates during the assembly procedure. Moreover, they also partly compensate for the differential shrinkage due to the fabrication process, improving the quality of the devices. Details concerning the optimization of the injection molding process will be given in a later paragraph. The flow distribution network has separate flow paths for stimuli as well as culture medium and cells. All implemented features are fully symmetric in order to assure identical hydraulic

4. Design of devices and mold fabrication

resistances. Standard HPLC fittings, valves and capillaries are used for the connection of an external flow injection system.

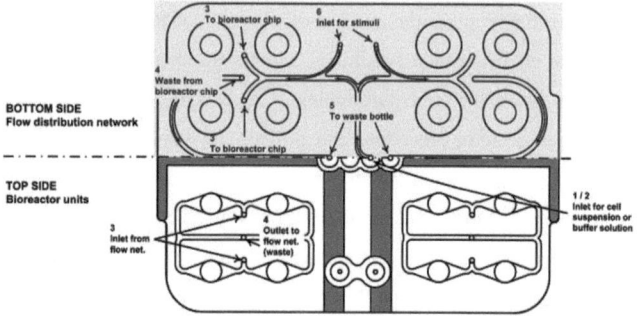

Figure 14. Biosensor array with embedded flow distribution network on the bottom side and four bioreactor units on the top side. Four QCRs (10 MHz, Ø 8mm) are fixed to the bottom side of each bioreactor unit with paraffin. The channel cross section is 1.0 x 0.8 mm, as a compromise between maximum tolerable shear stress and minimum stimulus consumption.

4.1.1. COMSOL simulations

The flow velocity profile in the bioreactor unit was simulated using COMSOL Multiphysics. The main focus here was the evaluation of how asymmetries due to fabrication accuracy can influence the velocity field distribution within the channel network. This was done in order to further validate the hypothesis of identical flow conditions on the surface of each QCR. A typical value of tolerance that occurs during fabrication of the molding form is ± 0.05 mm per 10 mm, which reduces the tolerances for smaller sizes. For a channel cross section of 1 mm x 0.8 mm a reasonable value of ± 0.025 mm was selected. Also the estimation of the shear forces acting on the cell layer was of interest. In fact, MDCK cells revealed to be relatively sensitive to shear forces [55]. Therefore, a second simulation was performed in order to exclude that the cell layer could be under stress due to the flow-through regime applied to the biosensor array during cell proliferation. In order to effectively include the influence of the fabrication accuracy, the computational fluid dynamics (CFD) model of the microfluidic network that was implemented in COMSOL Multiphysics, had to couple three different application modes, simultaneously affecting each other. The application modes which were employed are the "Solid, Stress-Strain", the "Incompressible Navier-Stokes" and the "Moving Mesh". The latter was used in order to couple the mesh with the deformations of the microfluidic network, which were introduced through static distributed loads implemented

4. Design of devices and mold fabrication

with the "Solid, Stress-Strain" application mode. The Incompressible Navier-Stokes module was then coupled to the moving mesh so that the simulated flow profile would reflect any deformation of the geometry. Furthermore, in order to optimize the memory usage during simulation, symmetry boundaries were introduced within the CFD model so that only a fraction of the whole microfluidic network had to be implemented. Figure 15a shows the model used for the simulation of the overall flow profile and estimation of the shear forces acting on the cell layer. Figure 15b shows the model implemented for evaluation of the influence of fabrication accuracy. Due to higher memory usage, the model had to be further reduced introducing additional symmetry boundary conditions.

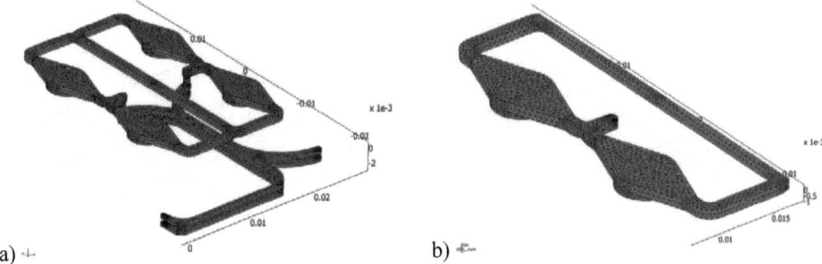

Figure 15. CFD models implemented in COMSOL Multiphysics for the simulation of the flow regime within the channel network. a) Model used for the evaluation of the velocity field and shear forces. b) Model used for the estimation of influence of fabrication accuracy.

The results of the simulation for the estimation of the shear forces on the cell layer are shown in figure 16. Figure 16a shows the slice plots of the simulated flow velocity profile and the estimated shear force [57] at 100 μm over the cell layer. In a reasonable range of flow rates up to several ml/h for a single bioreactor unit, the flow velocity is <<1 mm/s with a resulting maximum shear stress of about 13.3 μPa. For a comparison, the effects of shear stress on MDCK cells were previously investigated [56] showing morphological changes of the cytoskeleton starting from a shear stress of 9 mPa which is of about three orders of magnitude higher than calculated here. Also, preliminary experiments of cells cultivated in flasks placed on a shaker showed no significant differences in reference to a non-moving flask. Therefore, it can be stated that the influence of shear stress on the cell layer can be neglected for the range of flow rates here employed. With regard to the low flow rate the transport of species nearby the QCR due to convective mass transfer in addition to diffusion is comparably low. Figure 16b shows the plot of the velocity field along the dashed line at 100 μm over the cell layer.

4. Design of devices and mold fabrication

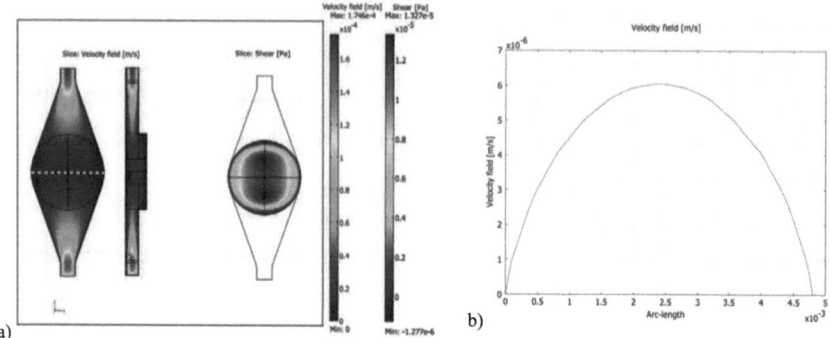

Figure 16. a) Detail of the simulated flow velocity profile and estimated shear force at 100 μm over the cell layer with a flow rate of 1 ml/h. b) Plot of the velocity field along the dashed line at 100 μm over the cell layer

The result of the simulation concerning the influence on the velocity field caused by asymmetries is shown in figure 17. Figure 17a shows the result of the simulation in the plane parallel to the sensor surface. Figure 17b shows the velocity field plot along the dashed lines, at 100 μm over the cell layer. The model takes into account asymmetries due to a fabrication accuracy of ± 0.025 mm. The variation of the velocity field values near the cell layer is less than 3% compared to an ideally symmetric device. Based on this result it can be assumed that all sensors are affected by the same flow regime.

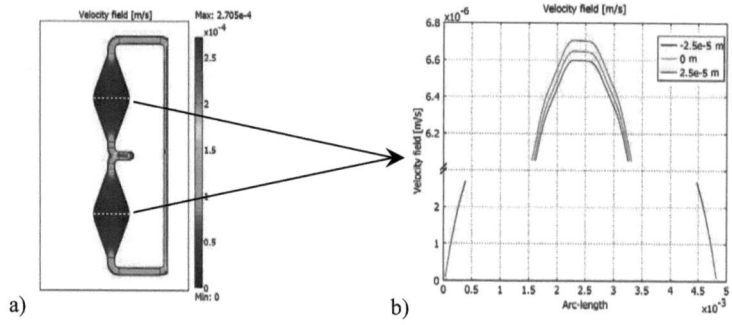

Figure 17. a) Simulated flow velocity profile within the micro-channels of the bioreactor unit with a flow rate of 1 ml/h. b) Plot of velocity field along marked red lines at 100 μm over the cell layer.

4.3. Molding process simulation and optimization

In order to optimize the geometries for the injection molding process, 3D models of the biosensor array were developed as basis for CFD models simulated using Moldflow code. The filling behaviour of the mold cavity, dependent on geometry, processing/material parameters and gate geometry/location were analyzed. For the final assembly of the micro-fluidic biosensor array it is very important to obtain parts with a flat surface and good dimensional accuracy. Therefore, the warpage and the volumetric shrinkage were chosen as criteria for part quality evaluation. It is known that polypropylene has a relatively large shrinkage factor which normally ranges between 1.2% and 2.2%. Therefore, it was very important to minimise this value as much as possible, especially in terms of the differential shrinkage that could generate stresses within the part and eventually deform or warp the flat surface. The first step of the simulation process was the calculation of the best gate location for the injection of the melt into the form. Next, an optimal gate geometry which could minimise shrinkage and warpage with a defined set of parameters was selected. Then, once the gate geometry has been chosen, the best processing parameters for minimum shrinkage and warpage were investigated.

4.3.1. Gate location

The simulation results to find the best gate location revealed that this is placed in the central area of the device. However, injecting the molten polymer from this area would mean to modify the geometry so that the functionality of the micro-fluidic network would be negatively affected. Therefore, the gate had to be placed on the side of the device which represented the second best choice for the location of the injection point. Figure 18 shows the top and the side view of the simulation results with the two considered gate locations.

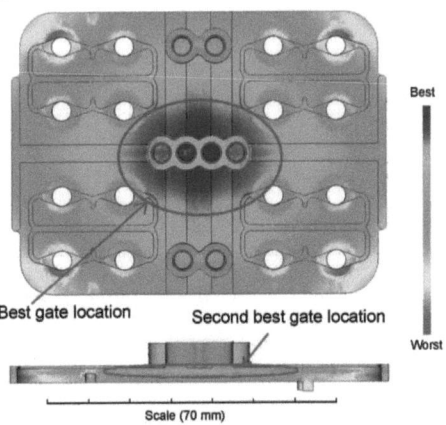

Figure 18. Top and side view of the gate location simulation results.

4. Design of devices and mold fabrication

4.3.2. Optimisation of the gate geometry

Three different CFD models with dissimilar gate geometries were implemented and simulated. The gate geometries which were considered are the flash gate, employed for fabrication of an early prototype of the device, the pinpoint gate and a triangular shaped gate. For the last two solutions, a hole has been placed in front of the injection point in order to induce the melt to spread into the form and avoid jetting problems during injection molding of the parts. Jetting occurs when polymer melt is pushed at a high velocity through restrictive areas into open, thicker areas, without forming contact with the mold wall. Jetting leads to part weakness, surface blemishes, and to a multiplicity of internal defects. The analysis of the filling behaviour of the mold, dependent on the three different gate geometries, was conducted by keeping constant the set of parameters, such as temperature, injection speed, pressure and clamping force, related to the fabrication process. The chosen set of parameters was the one used for the fabrication of the first prototype of the device. Figure 19 shows how the gate geometry can influence the shrinkage of the part leading to a more or less pronounced deformation respect to the desired shape. In figure 19, the simulation results relative to the deformation of the CFD model with triangular gate are shown. A magnification factor equal to 30 has been used for visualization of the deformed shape in figure 19a. Moreover, values of deflection and volumetric shrinkage calculated at 50 selected nodes of the mesh have been chosen in order to compare the results here obtained. The nodes are placed along 5 lines of 10 nodes each and they are shown in figure 19b. The data points are plotted from the leftmost to the rightmost point of the line. Figure 19c shows the graphs with the comprehensive data from all CFD models implemented. The colors employed recall the ones used in figure 19b so that each line can be recognized by referring to that specific color. In each case, the deflection is larger in the proximity of the borders of the device and it has a minimum in proximity of the centre. As well, the volumetric shrinkage is slightly higher near the edges but it does not show any minimum in the central area. However, the shrinkage has a very pronounced maximum near the injection point in the case of the flash gate and the pinpoint gate. Such maximum value is almost not expressed in the case of the triangular gate. The optimal gate geometry, which gives the minimum deformation and volumetric shrinkage, is the triangular gate. The maximum average volumetric shrinkage is obtained with the pinpoint gate while the maximum deformation comes from the flash gate geometry.

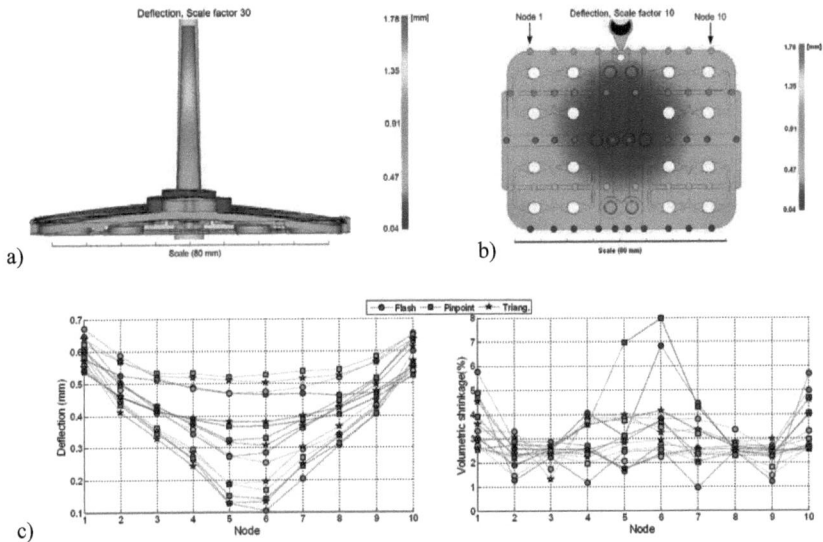

Figure 19. Simulated deformation and volumetric shrinkage of the molding injected biosensor array. a) Front view and b) top view of the deformed shape. The points highlighted correspond to the nodes of the mesh selected for the evaluation of the deflection and volumetric shrinkage. c) Plot for the comparison of the deflection and volumetric shrinkage values obtained from the simulation of all the three CFD models.

Based on these results, the gate with triangular geometry has been chosen for the fabrication of the molding injection form of the new device.

4.3.3. *Optimization of process parameters*

Injection molding process conditions must be accurately optimised in order to obtain parts of high quality. From the very beginning, it turns out that the potentially influential variables affecting the final result are numerous. Therefore, a wide set of parameters, such as injection time, injection speed, mold temperature, melt temperature, packing pressure, packing time and cooling time had to be considered during the optimization. Moreover, the processing variables do not influence the product quality separately but they extensively interact with each other and contribute to the result in different combinations. The process variables with the greatest impact on warpage and volumetric shrinkage are the mold temperature, the melt temperature, the injection time and the packing pressure over time [58, 59]. Regarding the warpage, the most influencing process variable is the packing pressure. Very important is also the contribution generated by the interaction between the mold temperature and melt temperature. The injection time, because of its short duration, has instead limited influence on

4. Design of devices and mold fabrication

the quality of the product. Nevertheless, in order to avoid differential cooling across the part that will induce warpage it is preferable to have the shortest possible injection time. On the other side, the shrinkage is mostly affected by the melt temperature. As well, the packing profile is of significance in order to acquire an even volume shrinkage. In this frame a constant packing pressure over time is to prefer in comparison with a linear increasing or decreasing packing profile. A compromise between processing parameters needed to be found in order to obtain both minimum shrinkage and warpage. The optimal injection profile selected is shown in table 1. Table 2 shows the setting for the packing phase of the injection molding process. The melt and the mold temperatures were respectively set at 230 °C and 40 °C with a cooling time of 40s.

Table I: Optimised injection profile for the fabrication process of the disposable devices

Injection profile	Step 1	Step 2	Step 3	Step 4	Step 5
Injected volume [cm^3]	10	4	4	4	1
Flow rate [cm^3/s]	30	25	20	15	10

Table II: Optimised packing profile for the fabrication process of the disposable devices

Packing profile	Step 1	Step 2	Step 3
Time[s]	0.5	1	18
Pressure [bar]	150	125	75

Figure 20a shows the simulated residence time distribution of the melt flowing into the mold (with triangular gate) of the biosensor array. All cavities could be filled without air traps and minimum shrinkage to achieve the required fabrication accuracy. A standard injection molding machine from Arburg (Allrounder 320S) was employed for the fabrication of devices made of medical grade of polypropylene. Figure 20b shows the new fabricated biosensor array.

4. Design of devices and mold fabrication

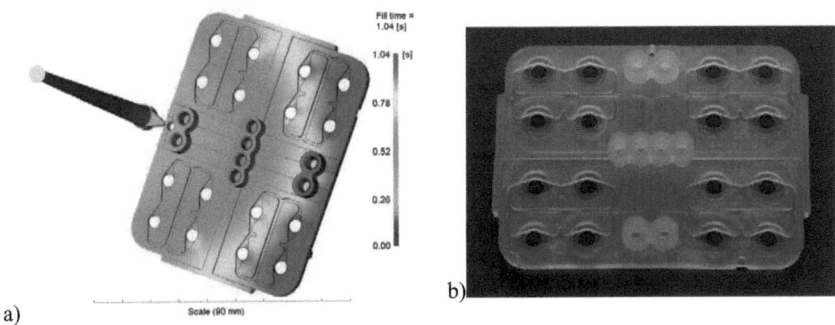

Figure 20. a) Simulated residence time distribution of the melt flowing into the molding form and b) fabricated disposable biosensor array (b).

5. FABRICATION OF SENSORS AND MEASUREMENT TECHNOLOGY

The fabrication of the sensors along with their placement into the biosensor array for the implementation of one single measurement unit will be described. Different processes were compared for the fabrication of the sensors. Gold electrodes were deposited on blank quartz discs and glass chips with transparent ITO microelectrodes were fabricated out of ITO coated glass wafers. Concerning the deposition of gold electrodes on the quartz discs, the main issue was to achieve a good and reliable metal deposition on the edge of the crystal in order to provide electrical contacts on the same side of the QCR. Evaporation and magnetron sputtering were evaluated and a set of shadow masks with a holder was fabricated in order to produce 16 QCR simultaneously. Glass chips with ITO microelectrodes were fabricated out of ITO coated glass wafers. Laser ablation and wet etching were compared for the structuring of the ITO electrodes. Laser ablation represents a quick and flexible solution for the fabrication of prototypes during optimisation. However, wet etching provides a "cleaner" result and should be the choice for the fabrication of the ultimate devices. The design of the sensor electronics will then be described in terms of architecture and performance. A simple equivalent circuit of the analog interface will be introduced and used for the estimation of the parasitic components due to the circuitry. Also, simple rules will be suggested for dimensioning the reference resistor embedded in the electronics. For correct acquisition of impedance values, a standard calibration procedure with references was performed. Finally, preliminary measurements were carried out in order to characterize the overall performance of the sensor array in terms of sensitivity, drift and standard deviation of the acquired impedance values. As well, a model of the ITO microelectrodes is proposed for fitting the impedance data.

5.1. Layout and fabrication of QCRs' electrodes

A single quartz crystal consists of a disc of 8 mm diameter. A big electrode with connection to ground should be deposited on the side that will be in contact with liquid phase during the measurements. A smaller electrode carrying the RF signal is deposited on the other side of the disc. Both electrodes should be accessible from the same side of the crystal and more precisely, from the side of the working electrode as shown in figure 21. Therefore, it is very important to achieve a reliable metal deposition also on the edge of the quartz discs in order to run the connection to the ground electrode on the same side as the working electrode.

5. Fabrication of sensors and measurement technology

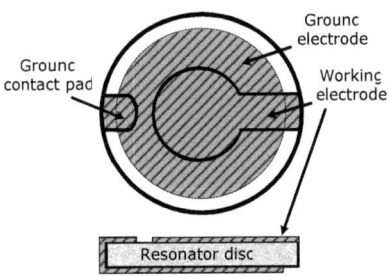

Figure 21 – Schematic of the quartz crystal resonator showing the electrode layout

Shadow masks have been fabricated for the simultaneous deposition of the desired electrode layout onto several quartz crystals. The masks are made by laser cutting (Nd:YAG laser with $\lambda = 1064$ nm) of 0.2 mm thick stainless steel sheets. Each mask is composed of three elements stacked on top of each other having the function of providing the shadow mask for the ground electrode, housing for the crystals and the shadow mask for the working electrode respectively. Four of these stacks are placed in a holder which keeps them together during the deposition process. The holder has been fabricated with the dimension of a 6 inches wafer. Figure 22a shows an exploded view of the assembly employed for the simultaneous electrode deposition on 16 quartz crystals. Figure 22b shows a picture of the fabricated masks with holder and deposited electrodes.

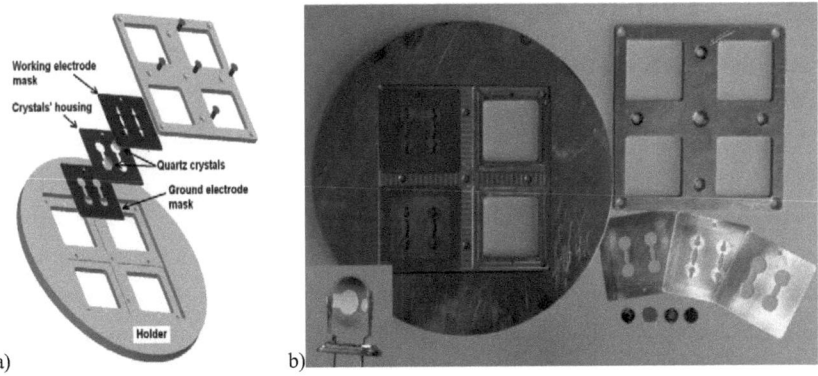

Figure 22 – a) Exploded view of the assembly employed for the simultaneous electrode deposition on 16 quartz crystals. b) Picture of the fabricated masks with holder and quartz crystals after electrode deposition.

Metal evaporation and magnetron sputtering techniques have been compared for the deposition of the metal. Evaporation produced electrodes with very sharp edges and well defined shapes. However, due to the absence of plasma, the deposition of the metal on the edge of the crystal could not be achieved. Sputtering suffered more of the planarity of the masks and in case of deformation some metal was deposited underneath the mask affecting the shape of the electrodes. However, the metal could be deposited on the edge of the discs in a reliable manner achieving an electrical connection with resistance less than 0.5 Ω over the edge. Therefore, magnetron sputtering was used for electrode deposition. Gold was used as the electrode substance. An under-layer of titanium was employed to promote good adhesion to the substrate. In order to evaluate the performance, some of the QCRs obtained were mounted on a standard holder and impedance spectra were acquired. Figure 23 shows the impedance spectra of a fabricated QCR having a minimum series resistance of 41.6 Ω and a quality factor of about 23900 in air.

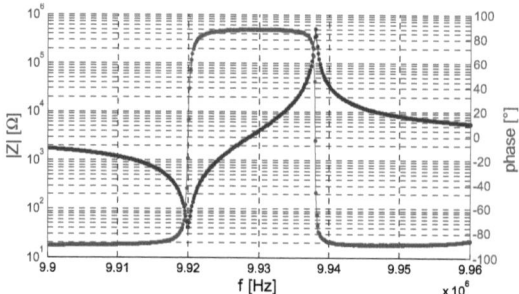

Figure 23 – Resonance impedance spectra with modulus and phase of a fabricated QCR in air.

5.2. Layout and fabrication of transparent ITO microelectrodes

Thin films of ITO (In_2O_3/SnO_2) deposited on glass are commonly used as transparent electrical conductors in touch panels, solar cells, liquid crystals or as coating for heat mirrors and electromagnetic shielding. ITO films can be deposited by chemical vapour deposition, sputtering or evaporation. The required patterned conductive electrodes are often achieved by means of chemical etching [62]. Chemical etching is relatively cheap. However, some issues like durability and adhesion of the photoresist in different chemical etchants, undercutting and areas of unetched ITO films can be involved. Moreover, the performance depends on the deposition method and on the Sn/In ratio of the ITO film. Also, the chemical etching involves the fabrication of masks which imply that changes to the layout can be achieved only through the realisation of a new mask, which represents an additional cost. Laser ablation can considered as an alternative for the patterning of ITO conductive electrode shapes [63 – 65]. It involves more expensive and bulky equipment but it provides higher degree of flexibility for

5. Fabrication of sensors and measurement technology

the purpose of fast prototyping of devices during the optimisation of the design. Different patterns can be quickly realised just by changing the file describing the geometry to implement. These two techniques were investigated and compared for the fabrication of the transparent ITO microelectrodes employed in this work. ITO coated glass wafers with a diameter of 100 mm were used in both cases. The nominal thickness of the ITO layer was 450 nm giving a resistance of 4 Ω/square. Figure 24a shows the geometry of the electrodes designed for the trans-epithelial impedance measurement. The round electrode in the centre is used as the working electrode. The second electrode, with the shape of a ring, has been introduced to be employed as guard electrode. Figure 24b shows the picture of a glass chip with ITO microelectrodes fabricated by wet etching process.

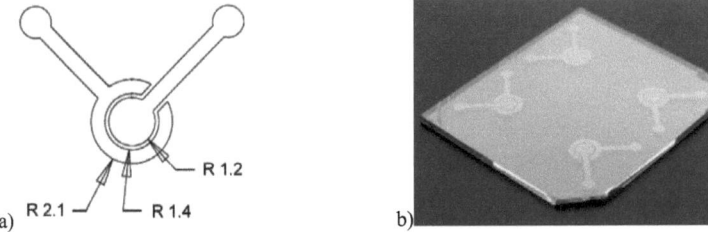

Figure 24 – a) Schematic showing the geometry of the ITO microelectrodes designed for the trans-epithelial impedance measurement. b) Picture of a glass chip with ITO microelectrodes fabricated by wet etching process.

5.2.1. Laser ablation of ITO layers

A pulsed Nd:YAG laser with wavelength $\lambda = 1064$ nm, maximum output power of 5 W and a spot size of 60 µm was used. The laser was focused perpendicularly on the ITO surface and a single scan was run for the removal of the deposited ITO layer. Table III presents a review of the laser parameters tested in order to find the optimum configuration.

Table III: laser parameters tested during the optimisation of the laser ablation process of the ITO layers.

Power [W]	Laser speed [mm/s]	Pulse frequency [kHz]
0.5	100	0 – 20 – 30 – 40
1	50 – 100 - 300	0 – 10 – 20 – 30 – 40
2	50 – 100 – 300	0 – 10 – 20 – 30
3	50 – 100 – 300	10 – 20 – 30
5	50 – 100 - 300	20 – 30

5. Fabrication of sensors and measurement technology

Criteria for the evaluation of the performance were the resistance value of the ITO, the quality of the edge created by the laser and the roughness of the glass in the areas where the laser has been focused. The best results were obtained at the speed of 100 mm/s with a pulse frequency of 20 kHz and power ranging from 2 to 3 W. Figure 25a shows a detail of the edge of an ITO layer after the laser ablation. The graph in figure 25b represents a profilometer scan across the edge of the structured layer. The plot reveals a smooth edge and a ridge protruding at the border of the ITO layer due to melting of the material. As well, the roughness of the substrate is increased of four times in the areas were the laser was focused. This is because the glass substrate partly absorbs the laser beam and gets damaged from the local heat generated.

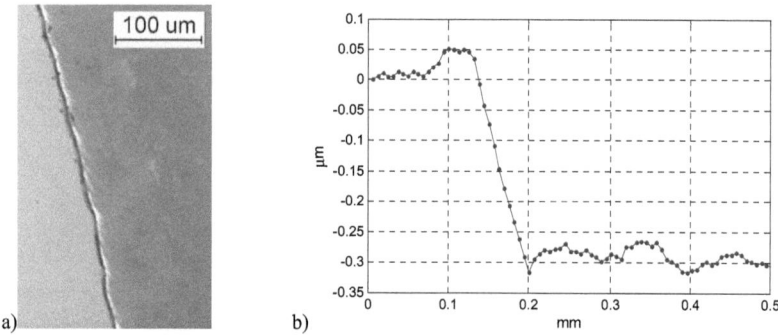

Figure 25 – a) Detail of a laser ablated area at the edge of the ITO layer. ITO: lighter area on the left of the picture. Glass substrate: darker area on the right. b) Plot obtained from a profilometer scan of the ITO layer across the edge.

5.2.2.   Wet etching procedure and mask fabrication

Etching of the electrodes was performed using 6 M HCl, 0.2 M $FeCl_3$ as etchant [62] and AZ1518 as photoresist. First tests of etching at room temperature showed an etching rate of 11.7 ± 1 nm/min. A mechanical profiler was used for the characterization. The temperature increase made the etching more effective. However, the edges of the ITO structures became "jagged" probably due to creeping of the etchant under the resist. Therefore, etching time was limited to 20-25 min at 50°C with a nominal thickness of 450 nm for the ITO layer. The resistance value and color of the etched ITO film were applied as control parameters during the final electrode fabrication. A 5 inches dark field chrome mask plate was produced by a private company and allowed the fabrication of four glass chips out of each glass wafer. Figure 26a shows the design used for the fabrication of the chrome mask. A picture of one fabricated microelectrode and a detail of the edge of the etched ITO layer are shown in figure 26b and figure 26c respectively.

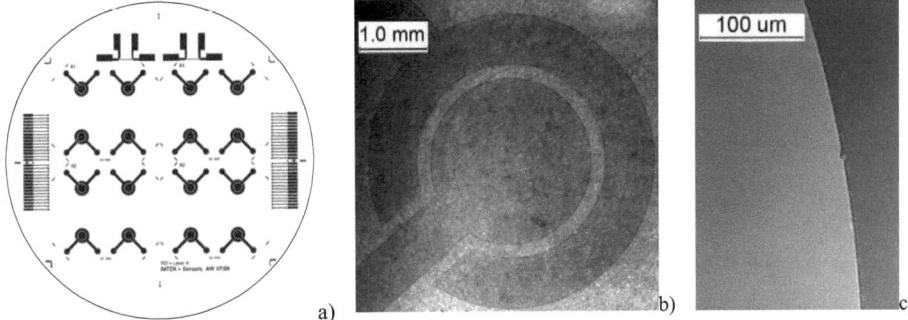

Figure 26 – a) Design of the chrome mask used during the wet etching process. b) Picture of one fabricated ITO microelectrode. c) Detail of the same ITO microelectrode showing the edge of the ITO layer. ITO: lighter area on the left of the picture. Glass substrate after etching: darker area on the right.

5.2.3. Comparison of laser ablation and wet etching

Pulsed Nd:YAG laser with wavelength $\lambda = 1064$ nm is suitable for fabrication of prototypes. However, it suffers of some disadvantages:

- At this wavelength the beam energy is partly absorbed by the glass substrate which gets slightly damaged during the process. Yet, other studies revealed that a better performance could be achieved with a laser with shorter wavelength [64]
- The laser patterning produced structures with somewhat more irregular edges than obtained with chemical etching. This is probably due to melting of the ITO layer during the ablation process. Also, melting of the ITO produced the formation of ridges protruding at the border of the layer.

These features are in general not desirable because they may be the source of distortion of the generated electric field. Therefore, chemical etching has been chosen for the fabrication of the final devices.

5.3. The measurement unit

Figure 27 shows how the QCRs and the glass chips with the ITO microelectrodes are embedded within the microfluidic device to implement one of the 16 measurement units of the biosensor array. The glass cover and the plastic parts are joined using a biocompatible epoxy resin whereas the QCRs are fixed in the biosensor array with paraffin. Using paraffin provides support to the QCRs along with effective sealing of the microfluidic network at the sensors positions without raising biocompatibility issues. Moreover, thanks to its low melting temperature, paraffin gives the possibility to achieve an effective but reversible mounting of the resonators which represent a very important feature in order to be able to retrieve the sensors before disposing the microfluidic device after use. However, this solution introduces additional damping on the QCRs. Therefore, special care is necessary to achieve proper mounting of the QCRs with the minimum amount of paraffin. Fig .27 shows in detail how QCRs are mounted within the biosensor units. The spring contact tips are used for electrical connection to the sensor interface electronics located under the array. The resonator ground electrode is always on top, in contact with the liquid phase, whereas the backsides as well as the spring pins are protected from these media via the paraffin sealing. O-rings are introduced in order to provide sealing against the overpressure which can be applied to the backside of all QCRs.

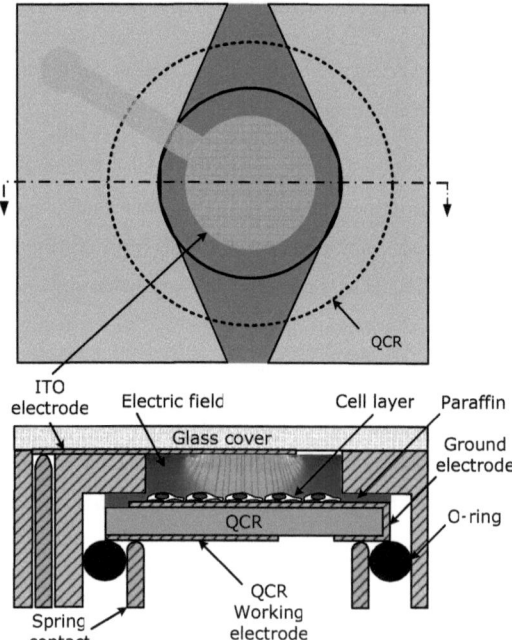

Figure 27 – Schematic with top view and cross section of a measurement unit of the biosensor array including QCR and ITO microelectrodes with spring contacts for electrical connection to RF and GND potential.

5.4. Measurement electronics

Sensor electronics consisting of an analog interface with an amplification stage and multiplexing capability were developed for the acquisition of the sensors impedance as a function of frequency with an in-house developed miniaturized impedance analyzer [16]. The multiplexer is controlled through the digital bits available on the impedance analyzer and it allows fast switching between the 16 QCRs and the 16 ITO microelectrodes along with selection of the relative reference impedance. Both wideband analog switches and a high frequency electromechanical relay were employed as a compromise to benefit from the advantages of the two solutions. Compared to electromechanical relays, analog switches are very small providing fast and bounceless operation with longer lifetime due to the absence of moving parts. Instead, electromechanical relays have a wider bandwidth; they provide almost ideal electrical connection when conducting, absence of crosstalk, no leakage currents, and comparably neglectable parasitic capacitance to the ground potential. Therefore, two sets of analog switches were implemented to achieve fast switching among the 16 QCRs and the 16 ITO respectively whereas an ultra-miniature relay from Teledyne has been employed for the selection between one of these two groups of sensors along with its reference impedance. The maximum switching time is given by the time necessary to switch between the two groups of sensors and it has a value of approximately 4 ms. This delay is not significant compared to the time needed for impedance spectrum acquisition (~1 s/ spectrum). Moreover, thanks to the synchronization between the digital outputs and the spectra acquisition it does not imply any loss of data when switching during the measurement. A block schematic of the developed electronics is shown in figure 28a. The circuit was optimized for fast switching and impedance spectra acquisition in a frequency range between 1 and 100 MHz. The electrical circuit in figure 28b was used as a model for characterization of the printed circuit board (PCB) and estimation of parasitic during the optimization of the board. Z_R and Z_{QCR} are respectively the reference resistor for the impedance measurement and the electrical impedance of the QCR connected to the interface. C_g is the parasitic capacitance to ground introduced by the input stage of the amplifiers, the multiplexer and by wiring. C_c is due to routing of the connections and the layout on the PCB.

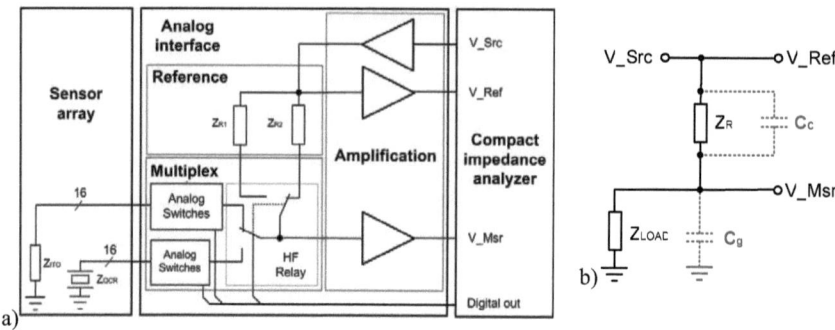

Figure 28 a) Sensor interface electronics for the biosensor array with multiplexer (16x) and analog input stage. b) Model of the analog interface used for the estimation of the parasitic components.

The estimation of the values for the parasitic capacitors was carried out by fitting the model with the data acquired from measurements with open circuit and short circuit load. Values equal to 85 pf and 0.9 pf were obtained respectively for C_g and C_c. The influence of C_c is small in the frequency range of interest and it has little influence on the overall performance of the electronics. C_g instead, has a significantly bigger value which considerably affects the measurement by reducing the bandwidth of the circuit. Indeed, (neglecting C_c) Z_R and C_g form a low pass filter between V_{Src} and V_{Msr}. Therefore, once the value of the parasitic capacitance C_g is known and the maximum frequency f_h at which to measure is given, the value of the reference resistor Z_R should be selected as follow:

$$Z_R \leq \frac{1}{2\pi C_g f_h} \tag{48}$$

Moreover, it has been shown [66] that the measurement error can be minimized by choosing the reference resistor:

$$Z_R = \sqrt{2} R_{LOAD} \tag{49}$$

Therefore, a compromise between (48) and (49) should be achieved when measuring a big impedance value over a wide frequency range.

5.4.1. Impedance analysis

Impedance spectra were acquired during experiments with the miniaturized impedance analyzer (MIA). However, also a commercial network analyzer system has been initially employed for comparison (Hewlett Packard 4395A, 801 data points per spectrum, [67]). Both systems require a standard calibration procedure with references. Therefore, the measurement data after calibration are similar. However, the calibration routine of the commercial system is limited to 50 Ω, while the MIA has the advantage to allow a reference-calibration with arbitrary resistance values, which can be closer to the sensor's series resistance, reducing the measurement errors in case of higher loads. Raw measurement data was acquired and a three point offline-calibration was performed independently for each sensor as described in [60] for further compensation of the parasitic components and correct measurement of the impedance values. An additional fit algorithm (polynomial regression 7th order) was then applied to the QCR calibrated spectra in the vicinity of the admittance maximum in order to obtain a more precise read-out of the resonance frequency (f_{res}) and conductance maximum (G_{max}) [66] values. Averaging was instead applied to improve the signal to noise ratio when measuring the ITO microelectrodes impedance.

5.5. Characterization of the sensor array

In order to verify the correct functionality of the sensor array, test measurements with water-glycerol solutions of 0%, 5% and 10% v/v of glycerol in water [61] were done. Figure 29a shows the spectra with magnitude and phase of a QCR mounted in the biosensor array acquired in air and in water. The minimum series resistance in water is of about 1 kΩ. As a compromise between (48) and (49) the reference resistor of the electronic interface has been dimensioned to 390 Ω. Due to the paraffin, a relatively high damping is present in air. When measuring in water, however, a similar response is obtained with or without the presence of the paraffin sealing. Hence, this solution does not significantly compromise measurements performed in contact with liquid phase. Figure 29b shows a graph with the motional resistance shifts of the QCRs at the resonance frequency due the load of the water-glycerol solutions. The resistance shifts are relative to distilled water. The difference between the continuous line, representing the experimental data, and the dashed line, showing the calculated theoretical values, can be attributed to the paraffin layer which contributes with a small additional load on the QCRs.

5. Fabrication of sensors and measurement technology

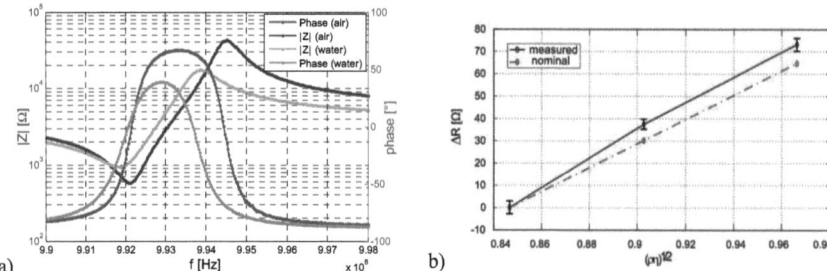

Figure 29 – a) Impedance spectra with magnitude and phase of a QCR mounted in the biosensor array in air and in water b) Measured and nominal values for the motional resistance shifts of the QCRs mounted in the biosensor array. The nominal values are calculated as a function of ρη. All values are relative to distilled water. Error bars refer to repeated experiments.

ITO microelectrode spectra were also acquired in parallel. The value of the reference resistor mounted in the analog interface was selected for measurements in highly conductive solutions. Therefore, changes in glycerol concentration were not detectable by this configuration. However, further measurements were performed using different conductivity calibration solutions and the media used for the cultivation of MDCK cells. Figure 30a shows the raw measurement data acquired with the MIA. The real and the imaginary part of the voltage divider ratio V_{Msr}/V_{Src} obtained using a reference resistor of 300Ω are plotted. The measurements show that the cell culture medium containing 10% of fetal calf serum (10% FCS) employed during the cell proliferation has a conductivity value half way between the solution with conductivity 12880 μS/cm and phosphate buffer saline solution (PBS, 15800 μS/cm), giving a value of 14300 μS/cm. Also, the real part of the voltage divider ratio had a value of about 0.31 at low frequency when measuring on 10% FCS. Reducing the value of the reference resistor to 200Ω increased the ratio to a value of 0.42 improving the sensitivity for measurements on similar loads. Following measurements on the ITO microelectrodes were performed using a 200Ω reference resistor, which is consistent with Eq. 48 and Eq.49. The voltage applied had amplitude of 80 mV. Although applying smaller voltages would minimize non-linear effects, an 80 mV amplitude excitation was preferred because of the better signal to noise ratio. Also, this value is within the upper range of naturally occurring membrane potentials. Yet, it has proven to have no significant influence on the electrical parameters of the MDCK cell monolayer [54]. Figure 30b shows a spectrum acquired from the ITO microelectrodes in 10% FCS (calibrated data) in the frequency range 100 kHz – 10 MHz, defined for the cell experiments. The graph shows a load which is mainly resistive ranging from a minimum value of 135 Ω at 350 kHz to a maximum value of 139 Ω at 10MHz.

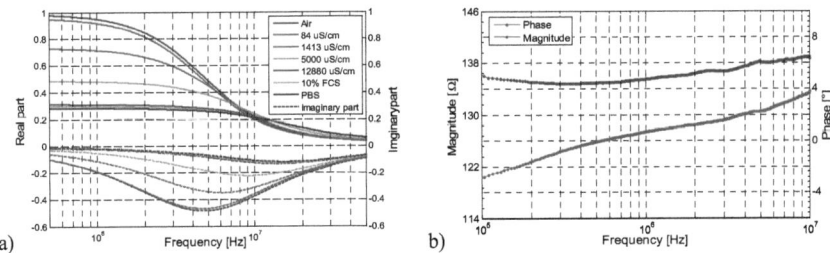

Figure 30 – a) real and the imaginary part of the voltage divider ratio V_{Msr}/V_{Src} acquired from one ITO microelectrode for different conductivity calibration solutions and for the media used for the cultivation of MDCK cells. b) Spectrum with calibrated data acquired with 10% FCS.

An equivalent electrical circuit was also developed for fitting the spectra measured with the ITO microelectrodes. PBS and cell culture medium were used in this frame. Figure 31a shows a schematic of the equivalent circuit employed. The calibration routine has been performed at the A-B plane. Therefore, the measured spectra are determined by the contribution of the ITO electrode in series with the electrolyte solution. Values of C_I and R_S were estimated in accordance with eq. (35) and (37) respectively. The value of R_{ct} was determined experimentally by measuring the DC current flowing through a big electrode placed in the solution. R_{ITO1} and R_{ITO2} are the values of DC resistance measured for the ITO electrodes between the points 1-2 and 2-3 respectively. The remaining parameters, L_{ITO1}, C_{ITO1}, L_{ITO2} and C_{ITO2}, were determined by a least squares fit of the measured data. With the exclusion of R_S, all the values obtained for PBS and cell culture medium are similar to each other. The contribution of the cell layer is not included in this model and will be explained in a later paragraph. Figure 31b shows the comparison between a measured impedance spectrum and the result of the fitting obtained using the model just described.

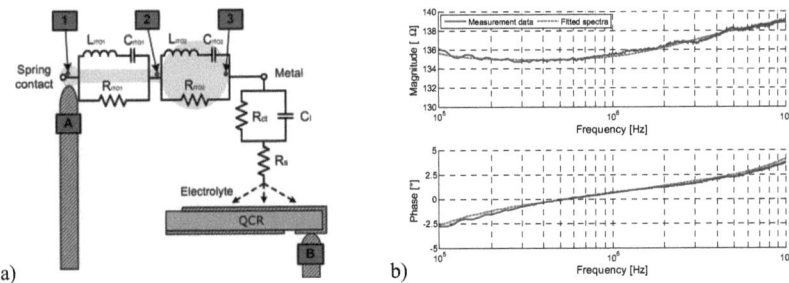

Figure 31 – a) Schematic of the equivalent circuit employed to fit the spectra acquired from the ITO microelectrodes. b) Plot representing magnitude and phase of a measured spectrum with the result of the fitting obtained with the model described.

Table IV summarizes the parameter values used to fit the spectra acquired in 10%FCS cell culture medium.

Table IV: Model parameters, fitted to the measured impedance spectrum depicted in figure 31b.

R_S [Ω]	R_{ct} [kΩ cm^2]	C_I [fF/µm^2]	R_{ITO1} [Ω]	R_{ITO2} [Ω]	C_{ITO1} [fF/µm^2]	C_{ITO2} [fF/µm^2]	L_{ITO1} [fH/µm^2]	L_{ITO2} [fH/µm^2]
133.8 ± 1	4.5 k ± 0.4	560 ± 60	31 ± 3	4 ± 0.3	177 ± 12	53 ± 2	6.2 ± 0.9	48.6 ± 5

5.5.1. Signal drift, precision and accuracy

Measurements over several hours in air, in PBS and in cell culture medium were performed to assess the stability of the measurement array as well as eventual drifts. The measurement system was placed into a custom made heating jacket connected to a thermostat for temperature control. Measured data for the QCRs and the ITO microelectrodes show that with a stability of ± 0.1 °C there is no actual need for temperature compensation. Concerning the QCRs, no significant drift was observed in air. A small resonant frequency drift of about 9 Hz/day along with a change in the minimum resistance of 3 Ω/day were detected in cell culture medium. However, the frequency and resistance values tend to stabilize over a time span of few days. As well, when flushing PBS afterwards an opposite drift with similar dynamics was detected. Therefore, absorption/desorption of medium components on the electrode surface were assumed to be the cause of this drift. A moving average with a 30 samples window was applied and a standard deviation of 0.4 Hz for the QCRs in air was determined. The value of standard deviation with liquid load was ranging from 2.5 to 9 Hz. The lower precision in water is clearly related to the higher damping of the sensor while its variability from sensor to sensor is associated to the paraffin mounting which was done manually. The readout of the magnitude value at the resonant frequency was less affected by circumstances related to the load or the assembly procedure, giving at all conditions a standard deviation of approximately 0.15 Ω. The ITO microelectrodes did not show any drift behaviour. Spectra could be acquired with a standard deviation of 0.055 Ω for the magnitude and 0.02° for the phase. Finally, to address the measurement accuracy, impedance spectra of the QCRs and the ITO microelectrodes were acquired with the Hewlett Packard 4395A impedance analyzer which was taken as a benchmark. Impedance magnitude and phase data obtained using the measurement array properly fit with the data obtained using the reference instrument with a relative error lower than 1%. Therefore, the measurement system implemented offers comparable measurement accuracy the Hewlett Packard 4395A.

6. SENSOR SYSTEM ARCHITECTURE

The present chapter will describe the full architecture of the biosensor. The microfluidic biosensor array along with the measurement unit composed by the sensor electronics and the miniaturized impedance analyzer has been discussed in the previous chapters. Therefore, the following paragraphs will focus on the description of the implemented microscopy pictures time lapse acquisition and of the flow injection system. Also, the implementation of an integrated shut off valve will be discussed at the end. This additional feature has the advantage to reduce the need of external HPLC equipment.

6.1. Overview of the measurement system

The microfluidic biosensor has been designed to be operated in flow-through and overpressure regime. An external flow injection system was used for automated and parallelized media feed as well as the control of the overpressure regime. This approach enables well defined proliferation and stimulation conditions. Besides, it prevents the accumulation of contaminants. Figure 32a shows an exploded view of the biosensor array including sensors, electrical connections, gaskets and fittings. This assembly is fixed on the PCB board through a special support with connection to the compressed air used to apply the overpressure. An external frame clamps the electronic board, the support and the microfluidic device together so that the o-rings are sufficiently compressed and can seal against the applied overpressure. The complete assembly is surrounded by a customized heating jacket connected to a thermostat for precise temperature control at 37°C. Figure 32b shows the fully assembled biosensor array with external equipment. A block schematic of the system with all connected devices is represented in figure 32c. The major components of the sensor system are the disposable biosensor array, the sensor interface electronics with a multiplexed analog stage, a miniaturized impedance analyzer, a flow injection system, a light stereo microscope (M205A) from Leica Microsystems and a nitrogen supply for overpressure control. With the exclusion of the thermostat, all devices were remotely controlled through a computer interface. The pressure regulation block consists of a manual regulation valve supported by a pressure sensor employed for fine adjustment and continuous monitoring of the pressure value.

6. Sensor system architecture

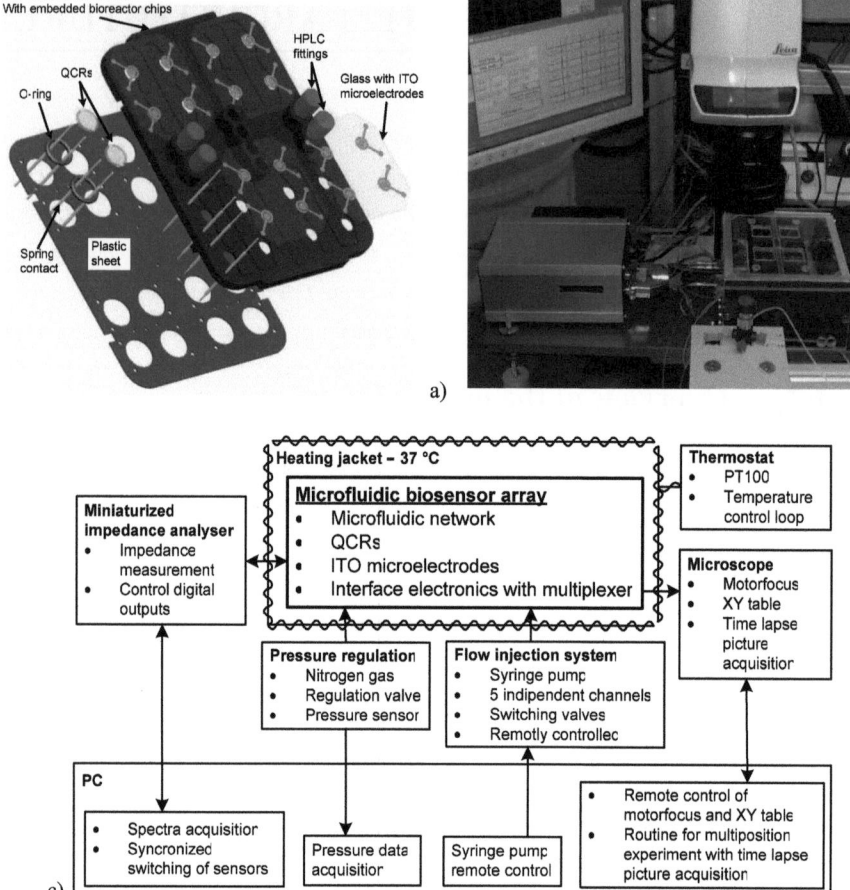

Figure 32 a) 3D model of the biosensor array. b) Micro fluidic biosensor array connected to the miniaturized impedance analyzer and placed in a heating jacket with transparent cover allowing for light microscopy observation of the sensor surfaces. c) Block schematic of the system implemented with all connected devices

6.2. Multi-position time lapse microscopy

Performing time lapse microscopy over a sensor array means to implement a "multi-position experiment loop". That is to say that besides acquiring pictures with defined time intervals there must be the possibility to constantly move the sensor array under the microscope objective (or vice versa) and acquire focused images for each sensor location. As well, the movements of the microscope should be synchronized with the picture acquisition. Therefore a stereo light microscope (model: M205A from Leica Microsystems) with motor-focus and a

XY table were employed for this purpose. Customized C++ software based on the Leica Abstract Hardware Model (AHM) [68] was developed. The purpose of the AHM is to provide a high-level access to microscopic hardware components. The model tries to give an abstract view on hardware parts and it is defined by a set of C++ header files, providing a set of virtual C++ classes (interfaces). In the AHM any connected hardware system can be exposed in a tree like structure where each node in the tree represents a hardware unit. The software provides control over all parts of the microscope, namely the zoom stage, the diaphragm, the motor focus and the position of the XY table under it. Moreover, it gives the possibility to save the status at 16 positions and to loop over these positions with a time interval which can be defined by the user. Pictures are acquired through a high resolution digital camera and proprietary software from Leica. At the beginning of the experiment, after the sensor positions are saved, the multi-position experiment loop is started and synchronized with the picture acquisition. Pictures taken from different positions are sorted and saved in different folders at the end of each experiment.

6.3. Flow injection system

The external flow injection system was implemented in order to provide continuous medium supply and parallelized stimulation of each independent bioreactor unit. Figure 33 shows a schematic with the main components of the flow system. A multi-channel syringe pump system from Cetoni (Germany) is attached for fast and well defined dispensing and medium feed. Dispensing of the culture medium during the proliferation process and the seeding of cells into each bioreactor unit can be respectively done through the inlets A and B. Furthermore, each unit is connected to dedicated inlets (S1-S4) for applying different stimuli in parallel. Multi-position valves at the inlets are used for flushing, cleaning and removal of residuals like gas bubbles and maintenance of the overpressure regime. A nitrogen bottle with pressure regulation valve is connected to the fluidic network as well as to the backside of all QCRs for applying a static overpressure. The reason for having overpressure in the system is that it prevents degassing of the medium and the consequent formation of gas bubbles in the channel network during long term experiments. Moreover, by this approach the same static pressure can be applied to both sides of the resonators avoiding bending of the quartz disc as well as mechanical instability. The connection between the external flow injection system and the micro fluidic biosensor array is achieved through standard HPLC equipment (tubing, valves and fittings). Thus, the fluidic part of the fully assembled sensor system follows the closed system architecture with no direct contact to gas phase.

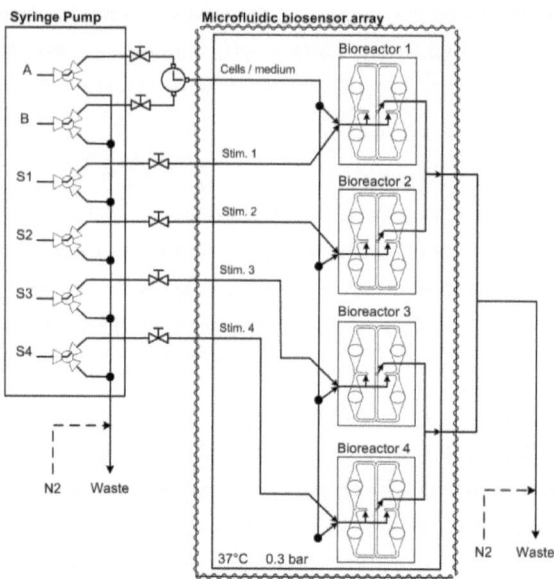

Figure 33 - Schematic of the flow injection system for culture medium feed and parallelized stimulation

6.3.1. Integrated shut off valves

The main idea of having a disposable device is to reduce time consuming washing procedures and decrease the risk of contamination. Ideally, everything that comes in contact with cells should be disposable. Unfortunately, HPLC equipment like fittings and valves cannot be considered disposable. Therefore, additional effort was given in order to integrate shut off valves within the microfluidic device and reduce the need of external components. Several kind of microvalves based on different working principles can be found in the literature [69]. However, their realization is often complicated and not suitable for disposables [70]. Yet, simpler realizations do not offer sufficient performance in terms of the displacement provided for the actuation of the valve [71] to be integrated into this microfluidic device. Therefore, a working principle similar to [72] was selected for the realization of a custom made pneumatically actuated shut off valve implemented using a thin silicone membrane. Figure 34a shows cross section of a shut off valve explaining the working principle. The liquid comes into a channel passing through a hole placed on top of it. A flexible membrane made of a thin sheet of silicone seals the channel from the bottom side. A plastic plate with holes placed at the same positions as the inlets made for the liquid is pressed against the membrane. When pressure Pv is applied through these holes the membrane deforms until it occludes the way to the liquid flowing in. As the pressure is released the membrane goes back to its original form and liquid can flow again. Thus, in this configuration the rubber sheet has not only the function of sealing the microfluidic channels, but it is also employed as an elastic membrane for the integration of pneumatically actuated shut off valves. Also, due to the phenomenon of force multiplication, which is obtained in hydraulic systems by altering the effective area on which a pressure is applied, the ratio $\varnothing v/\varnothing i$ should be selected >1. Therefore, the valve was implemented using values of 1.0 mm and 1.3 mm for $\varnothing i$ and $\varnothing v$ respectively which gave an amplification factor of 1.69 when considering the forces Fv and Fi originating from Pv and Pi respectively. Additionally, the channel geometry was especially modified at the inlets in order to provide additional room to facilitate the deformation of the membrane. The bottom view of the microfluidic device highlighting the areas modified for the implementation of the shut off valves is represented in figure 34b.

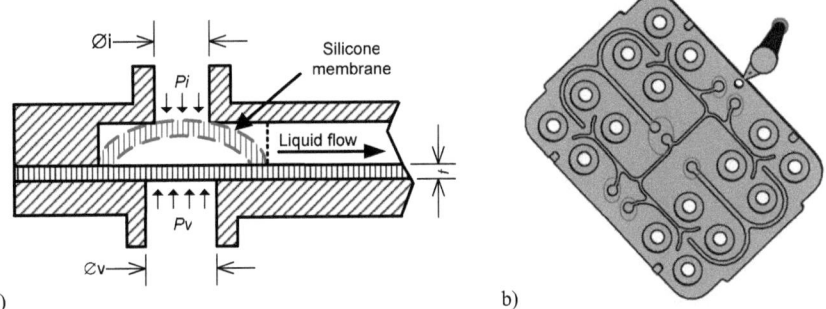

a)   b)

Figure 34 – a) Cross section of a shut off valve. The response of the valve depends on the dimensions $\varnothing v$, $\varnothing i$ and t. b) Bottom view of the microfluidic device. The areas modified for the implementation of the shut off valves are highlighted with red circles.

Since, the silicone rubber is permeable to N_2 gas (and to gasses in general) the actuation of the valve with compressed air would lead to the introduction of air bubbles into the microfluidic channels due to the applied pressure. As workaround, pressurized water was used instead of gas for the actuation of the valve. This was done because water has a bigger molecule than gasses and it does not permeate through the silicone lattice. Figure 35a shows a schematic of the fluidic connections realized to actuate the valve. The characterization of the valve was performed for two values of the pressure Pv. In this frame, the pressure Pv was set to a constant value and the liquid flow through the valve was measured for several values of pressure Pi. Figure 35b shows the characteristic curves of the valve obtained for Pv values of 2 and 3 bar. With an overpressure of 0.3 bar during cell culture experiments the working point of the valve would be placed at Pi = 1.3 bar. Hence, the curves represented in the graph reveal that Pv = 2 bar is suitable to be employed for the actuation of the valves integrated in the micro-fluidic biosensor array.

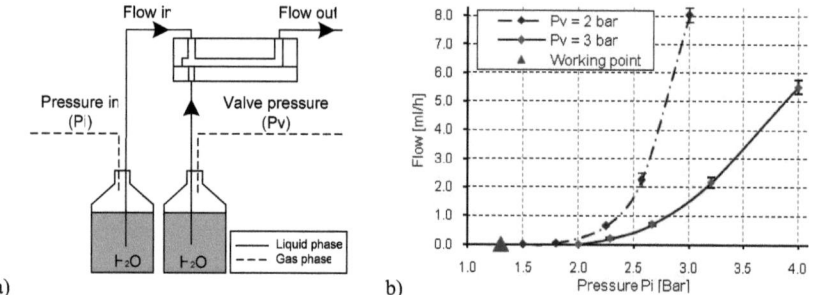

Figure 35 – a) Schematic of the fluidic connections realized to actuate the valve. b) Characteristic curves of the valve obtained for Pv = 2 bar and for Pv = 3 bar.

7. DESIGN OF EXPERIMENTS

The implementation of an appropriate protocol to apply during the experimental activities as well as the early interpretation of the experimental data acquired imply a stage of characterization of the measurement system in terms of handling issues. In fact, this is the early prototype a complex system combining several aspects. Therefore, it is important to dedicate additional time to the initial assessment in order to find out does and don'ts for the implementation of an appropriate experimental protocol. Hence, the following chapter will be dedicated to describe the main findings which lead to the implementation of an optimal experimental protocol. The sensor response to the liquid flow and to changes in the composition of media will be addressed. As well, issues related to the cell seeding procedure and to the cell stimulation will be presented. Finally, a detailed description of the optimised experimental procedure will be provided to the reader.

7.1. Response to liquid flow

As it will be explained in the following chapter, the flow rate of 1 ml/h has been selected to be employed for the cell proliferation. However, higher flow rates may need to be applied during some steps of an experiment e.g. for fast stimulation of the cell layer. Therefore, the response of the sensors to different flow rates was investigated. Flow rates ranging from 1 ml/h up to 16 ml/h were considered. In this range, the ITO microelectrodes revealed not to be sensitive to the liquid flow. However, the QCRs showed a response starting from 4 ml/h, especially in the resonant frequency value. The response is most likely due to a small bending of the crystal under the flow regime. This originates from the force acting on the crystal due to the pressure drop in the capillaries downstream the sensor. Figure 36a shows the resonant frequency shift of the QCRs with the different flow rates considered. A first small response was observed with flow rate of 4 ml/h. The resistance shift versus flow rate is represented in figure 36b. Here the first response could be observed from the flow rate of 8 ml/h whereas at 4 ml/h no resistance shifts were recognised. The entity of the error bars can be mostly accredited to the manual assembly of the QCRs with paraffin which introduces a term of variation in the response of the devices.

7. Design of experiments

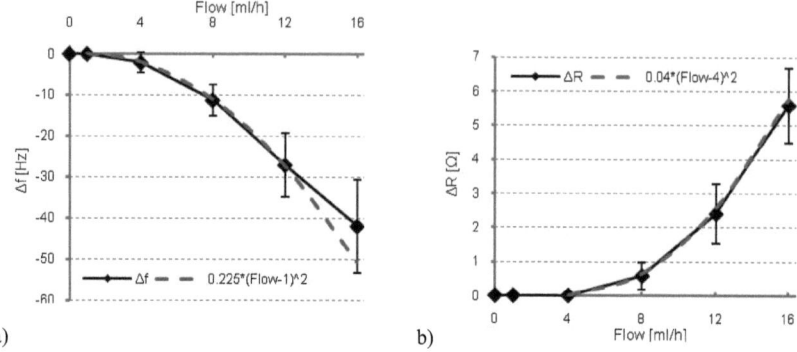

Figure 36 – a) Plot of the resonant frequency shift of the QCRs versus flow rate. b) Plot of the resistance minimum shift of the QCRs versus flow rate

Accordingly, the 6 ml/h was selected as the maximum value of the flow rate suitable to be employed during cell experiments. The value represents a good compromise between the need to provide quick stimulation and the necessity to eliminate measurement effects due to the flowing medium.

7.2. Changes in the composition of cell culture media

Preliminary measurements were done in order to verify the sensitivity of the measured ITO spectra to small changes in the composition of the media used during the experiments. Cell culture medium containing 10% of Fetal Calf Serum (10% FCS) was employed during cell proliferation. For the stimulation experiments, serum free medium was instead used. 10% FCS is obtained from 500 ml of Roswell Park Memorial Institute medium 1640 (RPMI 1640) by adding 50 ml FCS, 5 ml of penicillin/streptomycin and 10 ml of 1 M N-2-Hydroxyethylpiperazine-N'-2-Ethanesulfonic Acid (HEPES). Serum free medium comes from the same RPMI 1640 medium with the only addition of 10 ml of 1 M HEPES. RPMI 1640 is a basal medium consisting of vitamins, amino acids, salts, glucose, glutathione and a pH indicator. It contains no proteins or growth promoting agents. Therefore, it requires supplementation to be a "complete" medium. The reason of using serum free medium during the cell stimulation is to prevent unspecific carrier and capture effects leading to a decrease of biological active component concentration. This applies for example when stimulating with Hepatocyte Growth Factor (HGF) or when dispensing trypsin, an enzyme employed to induce cell detachment. The two different media compositions have comparable physical properties in terms density and viscosity. Therefore, QCRs could not distinguish them. As well,

7. Design of experiments

alteration in the media composition did not produce any measurement effect. Also the ITO microelectrodes were not able to distinguish between serum free medium and 10% FCS medium. However, they were sensitive to modification of the media composition due to cell metabolism. Serum free and 10% FCS media were aged by proliferation of MDCK-II cells for five days so that acidification would take place. Impedance spectra were acquired with fresh media and then with aged media flushed into the microfluidic biosensor array at the flow rate of 6 ml/h. Both media gave a similar read-out consisting of an increase of about 4.5 Ω in the magnitude at all frequencies and a neglectable phase shift of the impedance spectra in respect to fresh media. Aged media had a PH value of about 6.5 in comparison of a PH of about 7.5 for fresh media. This suggests that the ITO microelectrodes response may be dependent on the PH value of the solution [73]. Media taken from cell culture flasks after one day of proliferation could not be distinguished from the fresh media. Figure 37a and 37b show the change in the magnitude values of the measured ITO spectra at four selected frequencies after flushing of aged serum free medium and of aged 10% FCS medium respectively.

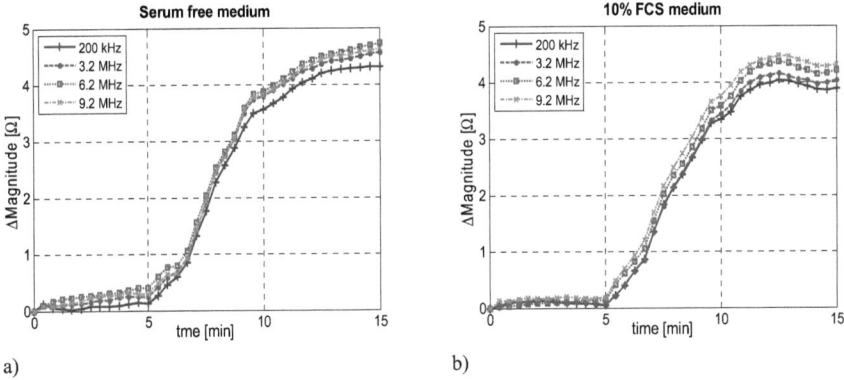

Figure 37 – Changes in the magnitude values of the measured ITO spectra at four selected frequencies after flushing of a) aged serum free medium and of b) aged 10% FCS medium. Aged media are flushed at minute five in both experiments.

The measurements done with the acidified media served to prove that the two kinds of media employed during the experiments with cells have the same properties in terms of the measured spectra. This information was an essential achievement in order to exclude measurement effects due to changes in the properties of the solution when exchanging from 10% FCS medium to serum free medium e.g. during cell stimulation experiments. Also, due to the constant flow through regime applied during the cell proliferation/stimulation experiments acidification of the media is normally assumed not to take place.

7.3. Cell seeding procedure

In disagreement with the assumption done in relation to the symmetrical structure of the microfluidic network, inhomogeneous distribution of cells was initially observed after the seeding procedure. Therefore, optimisation of the protocol applied during the seeding of cells into the biosensor array needed to be performed in order to satisfy the claim that cells are equally distributed on the surface of all QCRs at the beginning of each experiment. Experiments with ink visually confirmed that the flow is uniformly distributed throughout the microfluidic device. Thus, the reason for the unexpected phenomenon observed is to find elsewhere. The explanation can be found in the fact that after cells have been harvested and suspended they quickly start to attach to the capillaries of the flow injection system and to each other forming clogs. Also, due to the small size of the capillaries cells quickly deposit driven by gravitational forces and accumulate in short time at the bottom. These two facts together make so that if the cell suspension is flushed into the capillaries at a too slow flow rate or the flow stops even for a short time the cells are not evenly spread in the volume. As a consequence, when this suspension reaches a point where the flow divides in two channels the cells may not split equally and different numbers of cells reach different sensors. The outcome of the optimization is in agreement with these hypotheses. Indeed, it was found out that a relatively high flow rate (90 ml/h) is necessary to obtain repeatable and homogeneous cell coverage of all sensors in the array. Also, initially flushing a small volume of cell suspension to the waste and then switching to the biosensor array "runtime" significantly improved the reliability of the seeding procedure.

7.4. Stimulation of cells

Standard cell cultures are done in flasks, well plates or Petri dishes. In this context, the stimulation of the cells is normally done by adding a substance (or a mixture) in the cell culture medium or by exchanging the entire medium which was surrounding the cells during proliferation against another solution. This means that the stimulant is normally applied rapidly and then left in contact with the cells for the following time. By using coloured solutions, it has been estimated that a minimum dispensed volume of 1ml per bioreactor unit is necessary to exchange the liquid contained in the microfluidic channels. At the flow rate of 1 ml/h (the same as used during cell proliferation) it would take one hour to apply the stimulation to the cell layer. In order to be able to compare standard cell cultures done in flasks with the results obtained with the biosensor array, higher flow rates should be employed during the stimulation of cells. As well, the flow rate should not exceed a certain value not to induce any sensor response as seen in paragraph 7.1. Therefore, a volume of 1.5 ml dispensed at the flow rate of 6 ml/h has been employed as the initial step of all stimulations in order to achieve quick exchange of the medium present within the channels against the stimulus to be applied. As shown previously, this flow rate does not affect the resonator response extensively. Moreover, as it was proven with the COMSOL simulations introduced at paragraph 4.1.1 it does not imply any significant shear stress on the cell layer which could affect the cellular behaviour. Once the channels are filled with the solution containing the stimulant the flow rate can be set back to 1 ml/h for the remaining time of stimulation. Due to the overpressure regime, pre-compression of the syringe was necessary in order to avoid backflow of medium and reduce the risk of contamination from the waste bottle when applying a stimulus or dispensing media from a newly connected syringe.

7.5. Optimised experimental procedure

All micro fluidic channels were flushed several times with Mucasol 2.5% v/v and n-octyl-ß-glucopyranosid 0,1% for cleaning and sterilization after the assembly of the fluidic biosensor array and the flow injection system. Subsequently, the system was flushed for one hour with full culture medium (RPMI 1640 with 10% v/v Fetal Calf Serum, 1% v/v Penicillin/Streptomycin, 20 mM HEPES). At this frame point, also the sensor signal acquisition was started in order to acquire a stable baseline before cells could be transferred into the biosensor array. After being cultivated in conventional cell culture for two days, MDCK-II cells were harvested, counted and transferred into a plastic syringe with full culture medium. Cell counts of 0.4 Mio cells/ml, of 0.3 Mio cells/ml and 0.2 Mio cells/ml were used in order to obtain initial grades of confluence of ~50%, ~35% and ~20% respectively. During cell seeding, a volume of 1.5 ml of cell suspension dispensed with a flow rate of 90 ml/h provided homogeneous coverage of each QCR of the biosensor array. After seeding, the flow was stopped for a time of 15 min in order to let the cells settle down and start to attach to the gold surface of the QCRs. The height of the micro fluid channel above the QCRs is of only 1.3 mm. Therefore, the difference in settling time between cells nearby the glass plate and the sensor surface can be neglected. After waiting, the culture medium was dispensed by a syringe pump at the flow rate of 1 ml/h during cell cultivation. At this time point cells just settled and are very loosely attached to the gold surface. However, the flow rare applied is so small that it does not have the strength to move the cells away from their positions. Experiments to determine the impact of the flow rate on the cell proliferation will be presented in the following chapter. The cells seeded on the QCRs were simultaneously developing on each sensor of the array, which symmetry ensured a comparable confluency and identical proliferation conditions. A static overpressure of 0.3 bar was applied. During all experiments impedance spectra were acquired at the rate of 2 spectra/min/sensor. Microscopy pictures of all sensor surfaces were periodically acquired during the complete duration of each experiment. As already described, the measurement system was placed on the XY table of a fully automated microscope which could be controlled via computer interface. Thus, the location and the focus position of all sensors were saved in a file at the beginning of the experiments and then looped by the custom made positioning software. This allowed the implementation of a multi-position time lapse picture acquisition. Pictures were taken at the rate of 6 pictures/h/sensor during cell proliferation and stimulation. Stimulation was applied after a minimum proliferation time of 24 hours. At the end of each experiment the adherent cell layer was removed by means of Trypsin in RPMI 1640. The micro fluidic system was then washed with Mucasol 2.5% and filled again with full culture medium for the start of the next experiment.

8. CELL CULTURE EXPERIMENTS

This chapter will lead the reader through a selected set of results obtained in the course of a specific sequence of experiments designed to elucidate the influence of environmental parameters on the cells cultivated in the biosensor array. The experiments will be presented in different sections dedicated to exemplify how each parameter is reflected on the sensors' response. The first paragraph will present the normal cell proliferation and cell detachment followed by starvation experiments with non-flowing medium and with serum free medium (flowing). Then, the impact of the flow regime and of the cell distribution will be introduced followed by HGF stimulation experiments. Along with the results, every section will provide interpretations of the measured data for an understanding of the cellular behaviour. This, suggesting that the developed biosensor array could be successfully employed as a support to standard techniques for acquiring deeper insight into that complex system of communication that governs basic cellular activities and coordinates cell actions. In comparison with previous works, the possibility of comparing cell populations simultaneously proliferating in the biosensor array represented a significant step forward for the evaluation of cellular responses under the same experimental conditions. ITO microelectrodes were implemented into the sensor array in a later phase of the development. Therefore, only preliminary experiments for characterization of the response to the cell proliferation could be performed in a reasonable time before running the HGF stimulation.

8.1. Cell proliferation and cell detachment

Considering that MDCK-II cells can proliferate for three/four days in a T75 cell culture flask filled with approximately 15 ml 10% FCS medium, a quick estimation of the amount of media that should be flushed each hour through every measurement unit can be done by dividing these 15 ml by 84 hours (3.5 days). Considering that each bioreactor chip holds 4 measurement units fed in parallel, this leads to a need of about 0.7 ml/hr/bioreactor chip. Provided that this is a very rough estimation of the media consumption, a consistent safety margin was added to it and the flow rate of 1 ml/h was set for the first cell proliferation experiments. In this frame it was also assumed that the internal volume of the micro-fluidic device provides enough buffer capacity to ensure constant proliferation conditions for the cell layer. Later experiments, carried out to verify the impact of the flow regime on the cell proliferation, confirmed that 1 ml/h is an optimal parameter for the growth of MDCK-II cells in the biosensor array.

8. Cell culture experiments

8.1.1. QCRs response

Figure 38 shows the measurement data acquired from the QCRs of one bioreactor chip during a 2 days proliferation experiment along with microscopy pictures of the cells proliferating on a sensor electrode. The graph in figure 38a shows the comprehensive data from the experiment whereas figure 38b and 38c represent respectively zooms of the curve during the first five hours after seeding of cells and at the end of the experiment when trypsin was dispensed to induce cell detachment. Figure 38d shows a sequence of microscopy pictures taken from QCR4 with a time interval of 12 hours.

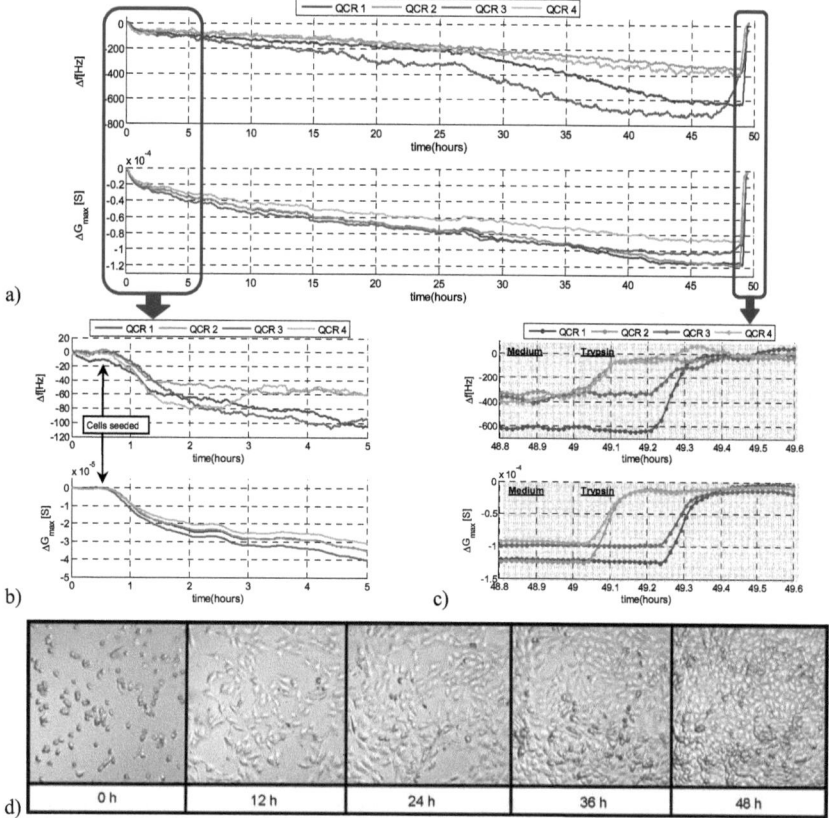

Figure 38 – a) Comprehensive data acquired from one bioreactor chip during a 2 days proliferation experiment. b) Zoom over the first five hours after seeding of cells. c) Zoom over the last 50 minutes of the experiment, during flushing of trypsin. d) Sequence of microscopy pictures taken with a time interval of 12 hours.

8. Cell culture experiments

Considering that cells proliferate over an area of 12.5 mm², several thousands of cells are randomly spread to cover the sensitive area of the QCRs. Hence, sensor signals provide an averaged read-out over the complete cell monolayer and a response of at least several hundreds of cells is required to observe significant changes in the measurement curves. Moreover, due to the constant proliferation condition guaranteed by the continuous medium feed it is assumed that only cell-covered parts of the surface contribute to the impedance spectrum shift. As well, dead cells or cell fragments floating in the bulk medium do not contribute to the sensor signal shifts. Negative shifts in the resonance frequency (Δf) and in the conductance maxima (ΔG_{max}) of the QCRs were observed already few minutes after the cells reached the sensor surface and started the adhesion process. Due to the initial cell attachment the greatest shifts were observed in the first hour after cell seeding. Sensor signal shifts due to cell division and accumulation are neglectable in this phase. Afterwards, the cell attachment can be considered completed and a slow and constant decrease of the measured quantities was recognised for each sensor. This is related due to an increasing acoustic load which, in agreement with microscope pictures, reflects the cell proliferation kinetics. A frequency deviation of 370 ± 183 Hz and a corresponding load resistance shift of 1006 ± 516 Ω were observed for confluent monolayers on a row of seven experiments. Results are somewhat similar to [6]. The big variation around the average values should not surprise. A cell culture is a complex biological system affected by many factors which are sometimes not fully understood. Even with the same nominal experimental conditions it is not possible to obtain 100% repeatability and each kind of experiment had to be performed several times in order to provide satisfactory statistics. Therefore, rather than only focusing on absolute values of frequency and resistance shift the analysis also considered the qualitative shape of the curves as well as the kinetics and the comparison of relative changes between the sensors of the array during a single experiment. Even though many of the details concerning the regulation of cell adhesion processes and the consequent sensor responses remain to be understood, four types of processes were here assumed to contribute to the QCR signals during the experiments:

- ➢ the initial cell binding to the sensor surface which implies the formation of focal contacts, leading to the first sensor response;
- ➢ the motility associated with the spreading of cells on the surface which increases the contact area leading to oscillations of the sensor signals;
- ➢ the modification of the adhesion properties along with number or type of binding proteins and strength of adhesion;

➢ modifications in the cytoskeleton of the cells which influence, for example, their rigidity and consequently have an effect on the conductance maximum (and possibly the resonant frequency).

To verify the effect of cell detachment trypsin was dispensed with a flow rate of 6 ml/h at the end of the experiments. Trypsin is an enzyme secreted in the duodenum which can hydrolyse the proteins responsible for forming the focal contacts between the cells and the sensor surface. Therefore, applying trypsin resulted in a relatively fast rounding up of cells and detachment from the sensor. As a result, the acoustic load on the sensor surface decreased quickly and an increase in the resonant frequency and conductance maximum could be observed within ten minutes with a phase shift of ~ 4 min in between. Both f_s and G_{max} went back to their initial values after ~20 min.

8.1.2. *ITO microelectrodes response*

The sensor response of the ITO microelectrodes during the cell proliferation was measured contemporaneously with the one of the QCRs. This provided significant additional information for the characterisation of the cell behaviour during growth. Spectra in the frequency range 200 kHz – 10 MHz were acquired with an occurrence of 72 spectra/hour. The spectra were then fitted by means of the electrical model obtained combining in series the models proposed in the previous paragraphs. All the parameters introduced at paragraph 5.5 and estimated for the measurement unit filled with 10% FCS without cells were kept constant. Only the values of the equivalent components C_1, C_2, R_1 and R_2 introduced (at paragraph 3.2.2.2) to describe the contribution of the cell layer were varied in order to fit the model for following the changes of the spectra during cell proliferation. Figure 39 shows the spectra acquired from the ITO microelectrodes during a proliferation experiment with a time interval of 12 hours between spectra. Two different representations are here represented. Plots of magnitude and phase versus frequency are shown in figure 39a while a representation in the complex plane is given in figure 39b. The dotted lines correspond to the spectra obtained from fitting the equivalent electric circuit with a least square algorithm to the measured data. The cell confluency on the sensor surface at the beginning of the experiment was ~ 30% and a confluent layer was formed after ~ 45 hours.

8. Cell culture experiments

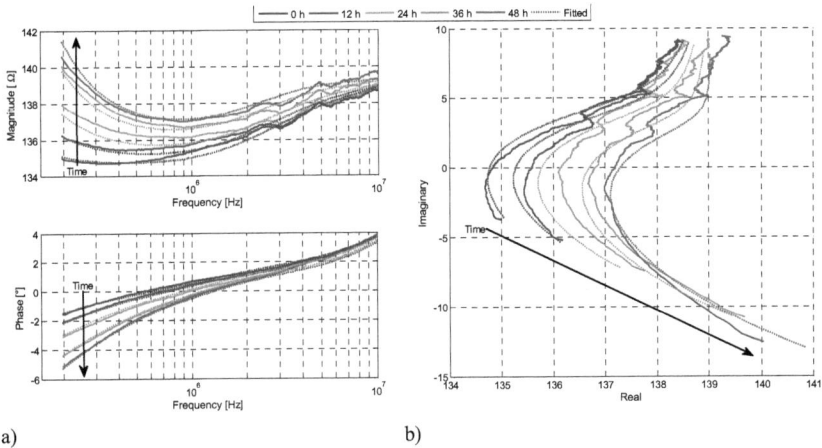

Figure 39 – Spectra acquired from the ITO microelectrodes during cell proliferation in two different representations. a) Plots of magnitude and phase versus frequency. b) Representation of the spectra in the complex plane. The spectra were measured during the same experiment which QCRs data is shown in figure 38.

Bearing in mind the complexity of the system being characterised, the result of fitting can be considered satisfactory. However, due to the partial electrode coverage the values obtained over time do not only account for the cell membrane resistance and capacitance but also for the fraction of free area on the electrode surface, namely the reciprocal of cell confluency. Therefore, an attempt to relate the confluency to the model parameters has been done. In this frame, the shifts of the spectrum occurring during cell proliferation at fixed frequencies and the values of the model parameters obtained from the fit were correlated to the grade of confluency observed with the microscope. As a result, a quadratic dependency between the electrode cell coverage and the model parameters has been found according to the following equations:

$$C_{1,2} = m_{1,2} \frac{electrode\_area}{confluency^2} \tag{50}$$

$$R_{1,2} = n_{1,2} \frac{confluency^2}{electrode\_area} \tag{51}$$

Figure 40 shows the shifts in magnitude and phase of the measured ITO microelectrodes spectra at seven selected frequencies during cell proliferation along with the comparison with the curves obtained from the equivalent electrical model. The curves are obtained from the spectra shown in figure 39.

8. Cell culture experiments

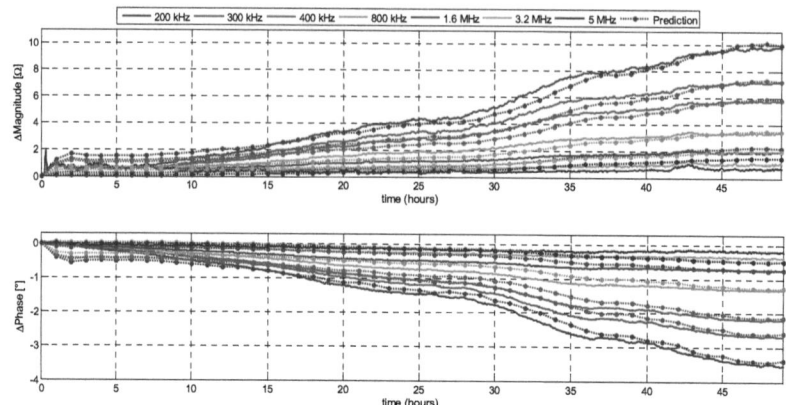

Figure 40 – Graph showing the magnitude and phase shifts of the ITO microelectrodes spectra at seven selected frequencies measured during cell proliferation. The dashed line represents the comparison with the curves obtained from the spectra calculated using the equivalent electrical model.

At confluency lower than ~30%, the curves extracted from the spectra calculated by feeding the electrical model with the parameters given by (50) and (51) do not closely follow the experimental data both in the magnitude and in the phase. However, the match between the model and the experimental curves significantly improves with increasing confluency. As well, the model gave the best correlation to the experimental data in the frequencies spanning from 200 kHz to 3MHz. The values obtained for $R_{1,2}$ were then fed into (46) and an estimation of the transepithelial resistance could be carried out by relating the so calculated R_0 to the area of the confluent cell layer. This led to a transepithelial resistance value of 153.6 ± 44.3 Ωcm^2 which is in agreement with values found elsewere [54, 95].

Concerning the cell proliferation kinetics analysis, three different phases distinguishing the cells behaviour could be recognised during each experiment. No sensor response could be related to the initial adhesion. During the first stage, called the lag phase, only a small step in the spectra could be observed at all frequencies right after cell seeding. There was little or no increase in cell number and the magnitude shift oscillated around a stable mean value during the following few hours. Nevertheless, cells were metabolically active increasing in size and showing some degree of motility. It is assumed that during this time, the cells are "conditioning" the media, undergoing internal cytoskeletal and enzyme changes and adjusting to the new environment. This phase had time duration of 5.7 ± 4.0 h. The cell distribution on the sensor surface after cell seeding is assumed to have considerable influence on the time extent of the lag phase. This was then followed by a proliferation phase during which the cell number increased constantly. At this point the spectra started to show increased shifts at

8. Cell culture experiments

frequencies lower than 1 MHz, whereas only smaller changes could be recognized in the higher frequency range. The growth continued as long as cells had free surface to support the increasing cell number. Due to the constant flow regime applied, nutrient depletion or waste accumulation were assumed not to be limiting factors for cell growth. During the final phase the number of cells remained constant and a plateau in the sensor signals could be observed. Over a set of seven experiments a magnitude shift equal to $9.1 \pm 1.5\ \Omega$ and a phase shift corresponding to $3.8 \pm 1.0\ °$ were obtained for confluent cell layers. At the end of the experiments trypsin was dispensed to induce cell detachment. As response, the phase went back to the initial value whereas the magnitude shift reduced to a value always lower than $2\ \Omega$ after some oscillations. Figure 41a shows the data obtained from a typical proliferation experiment with highlight on the three phases previously explained. A graph with a zoom on the first 12 hours of experiment is represented in figure 41b. Figure 41c show the zoom over the last 50 minutes of experiment when trypsin was dispensed.

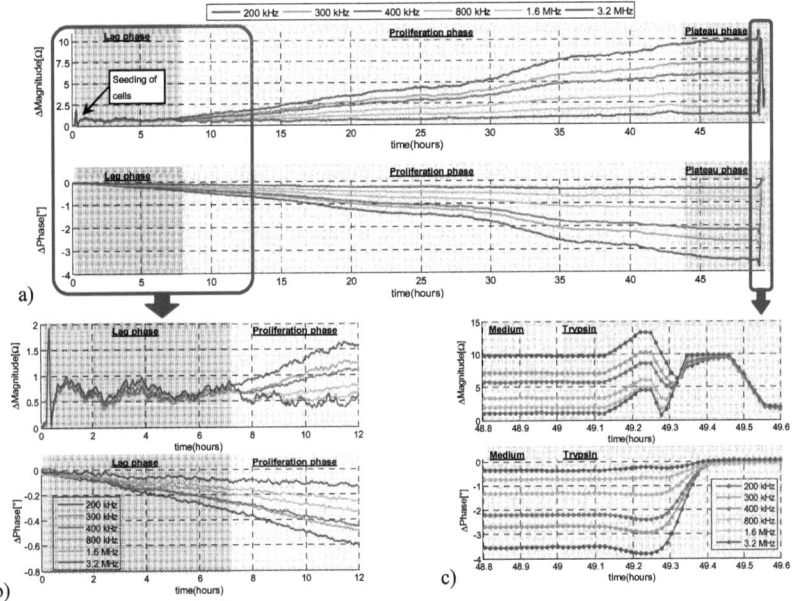

Figure 41 – a) Time plot of the shifts of the measured spectra at selected frequencies during a typical proliferation experiment. The three phases identified to distinguish different cell behavior are highlighted. b) Graph with a zoom on the first 12 hours of experiment showing the end of the lag phase and the beginning of the proliferation phase. The spike is due to the high flow rate applied during the seeding procedure. c) Zoom over the last 50 minutes of experiment during dispensing of trypsin.

8.2. Cell starvation

To verify the hypothesis that the internal volume of the device can provide enough buffer capacity cell starvation experiments with constant flow and with no flow were compared. For the stop-flow experiments, the flow of medium (10% FCS) was stopped after cells have been proliferating for 15 hours. The response was then compared with that obtained by exchanging 10% FCS medium against serum free medium in flow through regime. Figure 42a shows the result of the stop flow experiment. No response can be observed during the first hour. Then, a significant decrease in resonant frequency takes place during a period of roughly four hours suggesting a strong increase in adhesion strength of the cell monolayer due to a change in medium composition. This is followed by a phase of 10 to 15 hours where neither shifts in the measured impedance spectra nor significant changes in the cell morphology can be observed. After this, the resonant frequency and conductance maximum start to increase with a time shift of ~ 1 hour in between indicating the cell death. Figure 42b shows the outcome of experiment performed by exchanging the 10% FCS medium against the serum free medium in flow through regime. No response could be detected during the first 4 hours after the exchange of medium. Only a small oscillation of the conductance curve after 4-6 hours followed by a plateau could be observed. This, suggesting that no changes other than a gradual decrease in the proliferation rate of the cell layer can be attributed to nutriment starvation during the first hours after medium exchange.

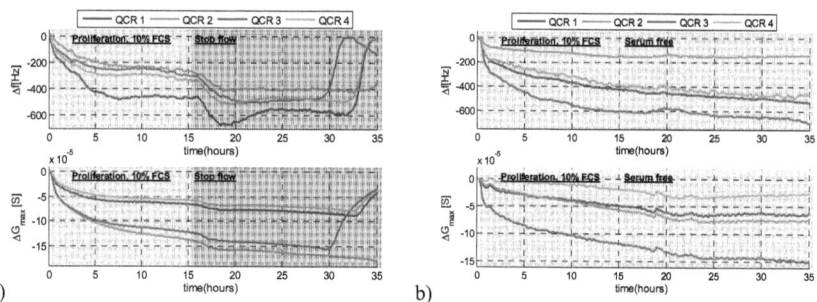

Figure 42 – a) Stop flow experiment: 10% FCS medium was flushed at 1 ml/h for the first 15 hours of proliferation after which the flow was stopped. b) Nutriment starvation: 10% FCS medium was flushed at 1 ml/h for the first 15 hours of proliferation after which serum free medium was replaced and flushed at 1 ml/h.

The comparison of the two experiments reveals that the changes observed in the sensor signal after the stop of flow are most likely due to accumulation of metabolic waste products rather than depletion of nutrients. As well, the buffer capacity provided by the 100 µl of medium surrounding the cell layer is sufficient to maintain constant proliferation conditions for about 1 hour after the stop of flow.

8.3. Impact of the flow regime on cell proliferation

The flow-regime nearby the sensor surface can have a major impact on the development of the cell culture. The COMSOL simulation results previously introduced revealed that the influence of the shear forces exerted on the cell layer by the liquid flow can be neglected. However, the convective mass transfer is regarded as an important factor for the cellular micro environment and hence for the measured cellular response on the QCRs. Therefore, the impact of the cell culture medium flowing in the closed micro-fluidic channels on the cell proliferation was analyzed. In this frame, MDCK-II cells were seeded into a single bioreactor chip and were cultivated at different culture medium flow rates in the range from 0.5 ml/h to 2.5 ml/h over several hours. For comparison cells were also cultivated in a conventional cell culture flask placed on a shaker located in an incubator. The setup was used to simulate medium flow on top of the adherent cell layer in conventional cell culture comparable to the situation in the biosensor array. The major difference is that the signaling proteins secreted into the extracellular medium remain in the flask, and cannot be washed out. The minimum cycle time which could be set for the shaker was approximately 9 s resulting in a slightly higher flow rate compared to the conditions in the bioreactor chip. Confluency, motility and morphology of cells were analyzed based on the light microscopy pictures acquired from the surface of all QCRs. When cultivating MDCK-II cells in a conventional cell culture flask on the shaker no significant differences in reference to a non-moving flask could be detected. Microscopy pictures showed a comparable cell morphology and cell division rate. Figure 43 shows the measurement data obtained from two experiments performed applying flow rates at the extremes of the range defined for this experimental phase. Figure 43a shows shifts in the series resonance frequency (Δf) and conductance maximum (ΔG_{max}) of 4 QCRs at the minimum flow rate applied, corresponding to total 0.5 ml/h for a single bioreactor chip. Two hours after seeding cells into the device sensor signals confirm normal adhesion of cells on the QCR compared to a flow rate of 1ml/h. In the following 10 h Δf exhibits a plateau or temporarily increases (QCR3), while ΔG_{max} decreases slightly. According to microscopy pictures, the plateau observed in Δf for QCR1, QCR2, QCR4 could be attributed to an almost constant number of cells adherent on the sensor surface, thus a minimum cell division rate. Figure 43b shows the cellular response at the highest flow rate applied, which corresponds to 2.5 ml/h. In the first two hours the adhesion and proliferation kinetics observed through the

8. Cell culture experiments

sensor signals and microscope pictures are comparable to what documented for the proliferation at 0.5 ml/h. In contrast after 3 to 5 h cells started to die and detach from the surface. Short term alterations of signals in both experiments can be mainly attributed to cell motility, while deviations in between QCRs are mainly due to a slightly different confluency and cell distribution on the sensor surface.

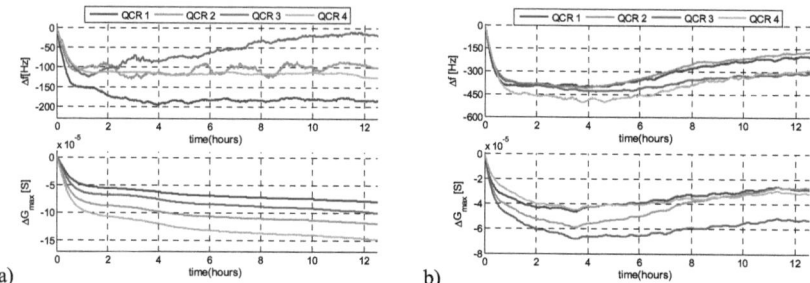

Figure 43 – a) Response of MDCK-II cells proliferating in a bioreactor chip at the culture medium flow rate of 0.5 ml/h and b) at the culture medium flow rate of 2.5 ml/h.

When comparing sensor signals with cell cultivation in flasks on a shaker, it can be concluded that fluid movement is not the governing factor for the cellular response at flow rates on the range of few tens of ml/h. Also, the COMSOL simulations results presented in a previous paragraph depicted that shear stress due to fluid flow does not play a role in the bioreactor chip. When decreasing the flow rate to 0.5 ml/h, first the cell division rate is reduced, but cells are still alive indicating insufficient yet not critical nutrient supply and removal of metabolic waste products. Conversely, when increasing the flow rate up to several ml/h, the convective mass transfer interferes with the cellular micro environment inducing cell death. With regard to the results from cell cultivation in flasks placed on a shaker, it is possible to assume that above a certain flow rate a washing out of species secreted in the extracellular medium due to convective mass transfer leads to cell dead. Therefore, the flow-regime and thus the flow rate has major impact e.g. on the cell division rate of MDCK-II cells. It can be stated that for the proliferation of MDCK-II cells in flow-through regime only a limited range of flow rates can be used for certain micro channel geometry.

8.4. Impact of cell distribution on proliferation kinetics

Concerning the large number (several thousands) of cells proliferating in the biosensor array, they can be considered as randomly distributed. However, dissimilar shifts in the electrical impedance spectrum can be attributed to differences in the cell distribution and hence a cell distribution dependent cellular response. In order to investigate this aspect, cell counts of 0.4 Mio cells/ml and 0.2 Mio cells/ml were used giving initial grade of confluence of ~50% and ~20% respectively. These two situations were employed to analyze to what extent differences in the initial confluency of the cell layer combined with the cell distribution may affect the cellular behavior in terms of the proliferation kinetics. Figure 44 shows the comparison of microscopy pictures obtained from two experiments where the two different initial cell counts were used. The pictures show the condition of the cells laying on the surface of each sensor shortly after the cell seeding procedure and after 20 hours of proliferation. In figure 44a and 44c the cell distributions on all 4 QCRs of a bioreactor chip, with initial cell counts of respectively 0.2 Mio cells/ml and 0.4 Mio cells/ml, are shown. Pictures of MDCK-II cells after 20 hours of proliferation, taken from the same areas, are shown in figure 44b and 44d. Although there is a similar confluency, significant variations in cell distributions on the sensor surface can be observed with the initial concentration of 0.2 Mio cells/ml. Also the comparison of the sensor surfaces after 20 hours of proliferation show significant differences. In fact, QCR 1 of figure 44b is almost confluent while the other three have lower confluency: QCR 2 and QCR 3 are almost identical to each other with the lowest amount of cells, whereas QCR 4 shows a condition in between the latter and QCR 1. On the other side, with the initial cell concentration of 0.4 Mio cells/ml differences on the initial cell distribution are less obvious to notice. At the beginning of the experiment (figure 44c), all sensor surfaces look similar and after 20 hours of proliferation all QCRs are covered with a confluent layer of cells. Such differences in the initial cell distribution led not only to changes in the absolute values of the measured signals, but also to differences in the proliferation kinetics between sensors in the case of lower initial cell concentration.

8. Cell culture experiments

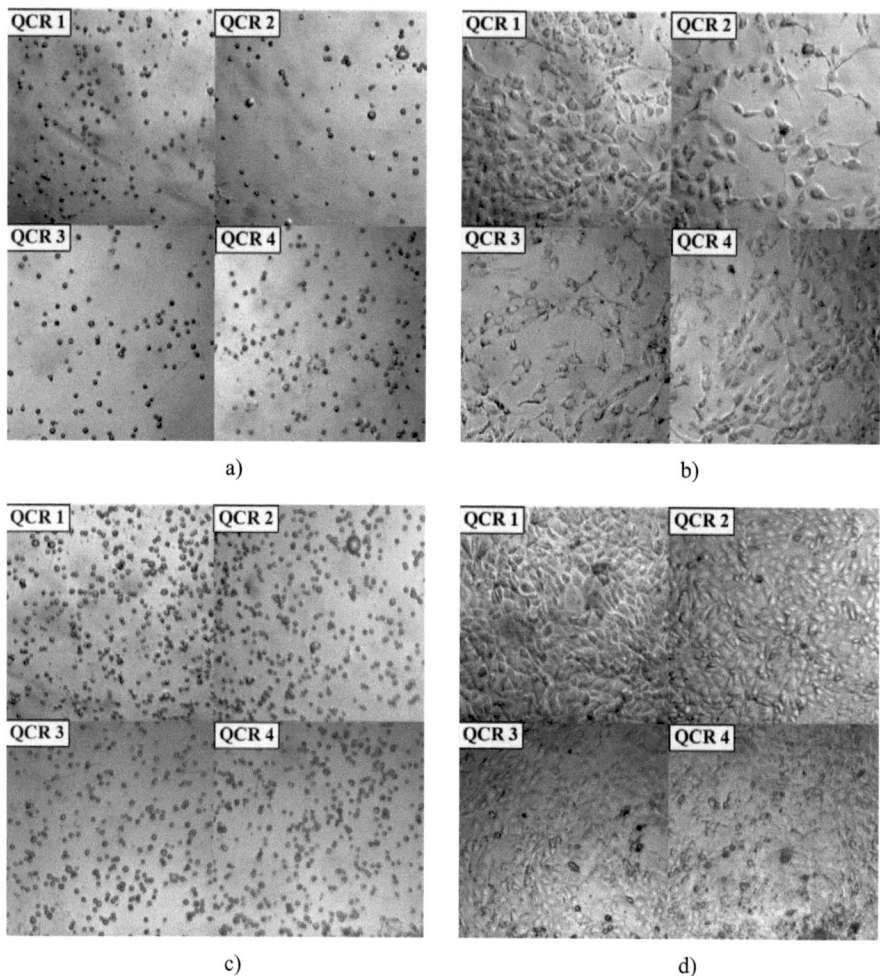

Figure 44 - MDCK-II cells distribution after seeding procedure (a and c) and after 20 hours of proliferation (b and d) on all four QCRs of the bioreactor chip with a) initial concentration of cells equal to 0.2 Mio cells/ml and c) 0.4 Mio cells/ml.

Figure 45 shows the resonance frequency shift of experiments with initial confluency of 20% (figure 45a) and 50% (figure 45b). Resonant frequency shifts relative to the value of the baseline acquired before seeding of cells are considered. During the first 30 min, a similar response consisting of fast decrease of resonance frequency can be observed in both curves. This is mainly due to the mass load of cells settling on the resonators and starting the adhesion process after the dispensing procedure. The lower shifts observed in graph a), in comparison with graph b), are due to a lower initial load on the QCRs related to the initial confluency.

8. Cell culture experiments

After the first hour, the proliferation process started and little differences could be observed between the chambers with 50% initial confluency. On the other hand, chambers with 20% initial confluency showed greater differences between the signals of the QCRs. In this case, the frequency shift curves also have different shapes. This is because the proliferation kinetics is not the same on the surface of each sensor. For instance, the series resonance frequency shift of QCR 3 in figure 45a differs in a way that after the initial decrease, due to cell seeding, it quickly reaches a plateau and it starts to significantly decrease only after 12 hours of cultivation. All the other QCRs instead, show a constant decrease from the very beginning of the cultivation. Also, for a given time point, relative changes in between absolute values are higher with 20% initial confluency than in the case with 50% initial confluency.

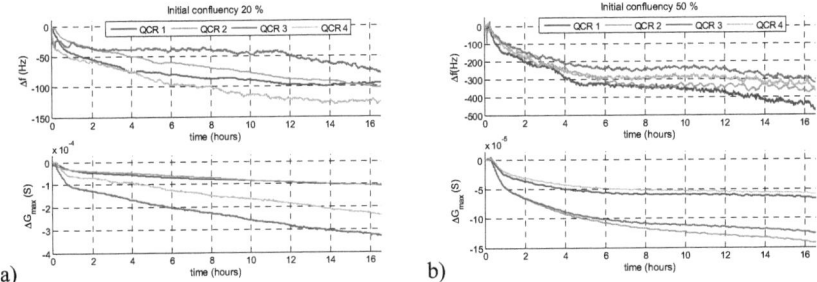

Figure 45 - Cell adhesion kinetics during proliferation of MDCK-II cells on QCRs with (a) 20% and (b) 50% initial confluency

To facilitate the comparison of the two situations, two different quantities were employed. In order to be able to quantify differences between the absolute values of resonance frequency and conductance maximum shifts a percentage variation between sensors was used. For a single experiment, this value was defined for each QCR as the frequency shift difference between the QCR being examined and a QCR chosen as reference, divided by the maximum frequency shift of the reference, as expressed in (52).

$$Variation\_i = 100 * \frac{\Delta f_{QCR\_i} - \Delta f_{QCR\_ref}}{\Delta f_{MAX_{QCR\_ref}}} \qquad (52)$$

Figure 46 shows the bar plots comparing the calculated percentage deviations among series resonance frequencies and conductance maxima in different experiments. The values are calculated taking QCR1 as the reference. Concerning the series resonance frequencies, the variation between the QCRs responses reached a maximum of 35% with the initial confluency of 50%. With the lower initial confluency instead the minimum variation was greater than 39% with a maximum of 50%. The variation between the values of conductance maxima was

8. Cell culture experiments

even more remarkable: it had values ranging from 13% to 40% for the experiments with the higher initial confluency and values almost three times as higher, ranging from 42% to 116%, for the experiments with the lower initial confluency.

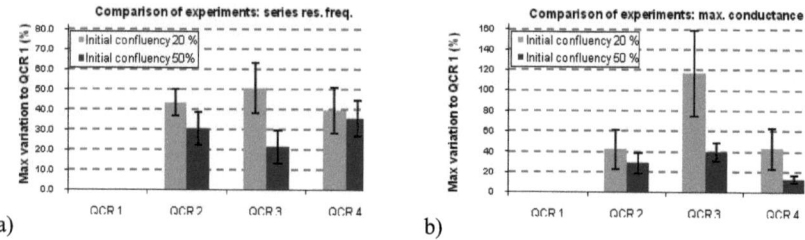

Figure 46 - Maximum variation of a) series resonance frequencies and b) conductance maxima between QCR2-4 referring to QCR1 averaged over 5 experiments.

For the comparison of the initial proliferation kinetics, a time constant based exponential behaviour was instead used:

$$Kinetics = \Delta f_{10}\left(1 - e^{-\frac{t}{\tau_k}}\right) \qquad (53)$$

where τ_k is the kinetics time constant and Δf_{10} is the value of the frequency shift after 10 hours of cultivation. A comparison between the data acquired during the cell cultivation experiments and the calculated exponential curves is shown in figure 47a. The comparison between the time constants is shown in figure 47b. In this case it is possible to observe that the average time constant calculated from the cultivation experiments with 20% initial confluency is more than twice the one obtained for the experiments with 50% initial confluency. Moreover, in the first case there is a much bigger variance from the average value which suggests again the higher influence of the distribution of cells on their behavior during the first hours of cultivation.

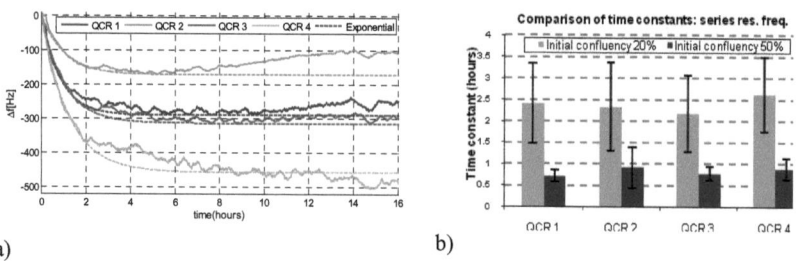

Figure 47 - a) Fitting of the experimental data with a time constant based exponential behavior. b) Comparison of the time constants calculated for the evaluation of the initial cell behavior during cultivation experiments.

8.5. Cell stimulation with Hepatocyte growth factor

In molecular cell biology the analysis of the interaction between HGF and epithelial cells is of special relevance. HGF is a stromal-derived multipotent polypeptide that regulates cell growth and motility of several cell lines, including MDCK-II cells. In fact, it has been discovered through its property to induce the growth of hepatocyte cell cultures but it also has the property to trigger scattering of epithelial cells [90, 91]. It mediates multiple biological responses including mitogenesis, motogenesis and morphogenesis in a wide range of tissues by activation of its associated tyrosine kinase receptor Methionine proto-oncogene (MET). Therefore, it plays an important role in tissue repair due to an increased proliferative activity and cell motility [89, 94]. The HGF-MET system is necessary for embryonic development. However, abnormal MET signaling contributes tumor genesis and metastasis [2, 3]. Although HGF is mostly known to have anti-apoptotic properties the MET receptor is remarkably able to promote apoptotic responses of few cell lines [93], which is again very important for the regulation of the survival/apoptosis balance during the development of tumors. By altering the cytoskeleton and morphology of cells, which is accomplished by a number of different proteins [88], HGF stimulation of MDCK-II cells triggers the formation of leading edge protrusions required for forward movement. Cell motility is also accompanied by a considerable decrease in number and size of stress fibers [90]. A decrease in cell-cell adhesion molecules and the subsequent loss of cell-cell contacts and cell motility can be observed at the microscope.

By triggering different structural changes within the cytoskeleton of the cell, HGF is likely to modify the acoustic load of a cell layer laying on a QCR. As well, the impedance spectra measured from the ITO microelectrodes were assumed to be affected by two factors combined with the HGF stimulation:

- the HGF mediated activation of membrane K^+ currents [94], leading to a change in the resistivity of the cell membrane;
- the decrease in cell-cell contacts produced by the cell scattering, leading to an increase in conductive paths through the cell layer.

Therefore, HGF stimulation surely represents an attractive stimulant to be applied to adherent cells growing into the microfluidic biosensor array for investigating the kinetics of HGF response mediated by changes in the acoustic load on QCRs and in the trans-epithelial impedance spectrum. Special focus was on the short term cellular response directly after HGF

8. Cell culture experiments

receptor triggering. In conventional cell culture experiments cell motility can firstly be observed on light microscope pictures after 1 h. Previous results published by the same group using a bioreactor unit with a single QCR showed that HGF involves at concentrations ≥10 ng/ml a change in acoustic shear wave propagation, respectively cell adhesion, on the sensor [12]. Here, the influence of two different HGF concentrations (80 ng/ml and 40ng/ml) and cell confluency at the time of stimulation (50% and 90%) has been investigated with the new biosensor array. Especially the comparison of sensor signal changes and optically observable cell motility were of interest. The cell count of 0.2 Mio cells/ml was used for all experiments. In analogy with the protocol applied in standard cell cultures, cells were let proliferate until they reached the desired confluency. Then, 10% FCS medium was exchanged against serum free medium. The lack of serum led to a slightly decreased proliferation rate. After 5 hours of equilibration HGF in serum free medium was dispensed. Flow through regime was applied during each step of the experimental procedure which enabled well defined stimulation conditions without critical accumulation of contaminants. When stimulating at 50% confluency, cells spread in a time scale of hours over the surface and the colony formation disappeared. With 40 ng/ml of HGF concentration, the observed cell motility was slower and less pronounced. When applying the 80 ng/ml HGF concentration cells spread much faster up to a plateau or saturation. When stimulating at 90% confluency, although cell spreading could not take place due to the lack of free surface, reduced cell-cell contacts and increased cell motility expressed in a quick rearrangement of the cell layer could be observed. Figure 48 shows a sequence of microscopy pictures with 2 hours interval in between illustrating the effect of HGF stimulation (80 ng/ml) performed on MDKC-II cells grown into the microfluidic biosensor array. The time ranges from two hours before until four hours after the stimulation. A significant increase in cell motility and consequent scatter of cells can be recognized already two hours after stimulation.

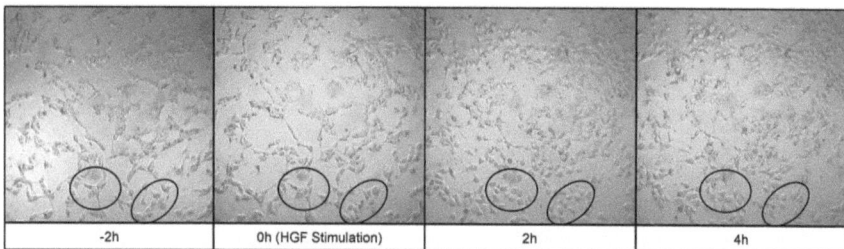

Figure 48 - Sequence of pictures showing the effect of HGF stimulation on MDKC-II into the microfluidic biosensor array. Two different cell colonies are highlighted for better comparison: cells show almost no motility during the two hours before stimulation whereas obvious scatter can be seen during the following hours.

8.5.1.   QCRs response

In a time scale of tens of minutes no response like changes in cell morphology or cell motility could be observed on microscope pictures taken from a conventional cell culture. Compared to the results in 24-well plates, cell scatter (on 50% confluency experiments) was less prominent in the micro-fluidic biosensor array, probably due to the reduced space. In a time frame of 1 h after stimulation with HGF only a small cellular response could be observed in microscopy pictures. More significant changes in cell morphology appeared >2 h after beginning of treatment with HGF as already shown in figure 48. Figure 49 shows the sensor signal during the first hours of HGF stimulation at 80 ng/ml concentration. Over a time period of ~25 min a slightly decrease in f_s and G_{max} was measured. Although cells were in contact with serum free culture medium, sensor signal changes could be mainly attributed to cell proliferation. Afterwards, the frequency shift curves started to increase with variable slope or flattened and increased with a delay of 1 to 2 hours. Between 2.5 and 3.5 hours after stimulation a local maximum was reached and then the curves decreased again as during normal cell proliferation. The conductance maximum curve showed a different behavior. After the first 25 min, it strongly decreased down to a local minimum before increasing to a local maximum and stabilizing after ~3.5 hours. Besides showing difference in the extent of frequency and conductance shifts, similar kinetics of HGF response were detected both with 50% confluency and 90% confluency.

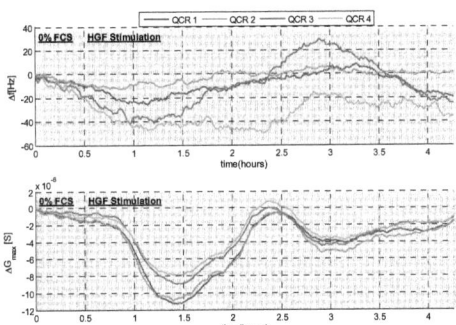

Figure 49 – QCRs' sensor signal shortly before and during the first four hours of HGF stimulation. Cell confluency: 50%. HGF concentration: 80 ng/ml. Shifts are calculated from the beginning of this time window in order to superimpose the signals and be able compare them at a glance.

A cellular response could be detected within a time of 30 min, even though no obvious visual changes in cell morphology as well as cell motility could be observed. This means that the

8. Cell culture experiments

acoustic load, respectively stiffness and density, of the cell–sensor interface changed significantly before the scatter took place. However, due to the sensitivity of the QCRs to the spatial distribution of the cell colonies, changing between experiments, a strict characterization of the sensor response in terms of frequency and conductance shifts could not be performed. Nevertheless, the oscillations observed in f_s and G_{max}, which are regarded as a decrease in adhesion strength combined with increased viscous behavior during cell movement, allowed the identification of a time window of ~4 hours for the main HGF response along with turning points at 30 min, 1 hour, 2 hours and 2.5 hours after stimulation.

8.5.2. *ITO microelectrodes response*

Already during cell proliferation experiments, the measurement signals obtained from ITO microelectrodes showed little sensitivity to changes induced by a cell layer with confluency lower than 30%. This can be related to the relatively high portion of free sensor area which introduces a low resistance path in parallel to the cell layer. Running HGF stimulation at 50% cell confluency provided a good read-out from the transepithelial impedance measurement. However, the HGF mediated sensor responses obtained when stimulating at 50% cell confluency were similar to what obtained at 90% confluency. This is valid for both the amplitude shifts and the kinetics of the response. Figure 50 shows the response of at 50% confluency cell layer stimulated with HGF concentration of 80 ng/ml. Responses to HGF concentration of 40 ng/ml are similar.

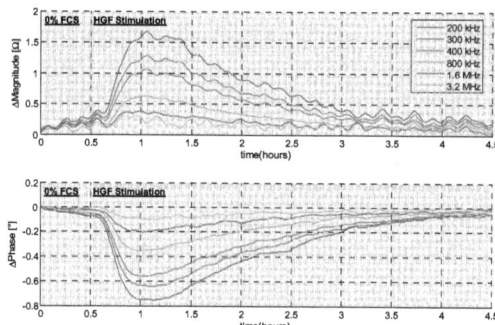

Figure 50 – ITO microelectrodes' sensor signal shortly before and during first hours of HGF stimulation. Cell confluency: 50%. HGF concentration: 80 ng/ml. Shifts are calculated from the beginning of this time window.

After 15 min from HGF stimulation, the magnitude of the spectrum starts to increase until a local maximum is reached after ~30 min. Then, the curves start to recover towards the initial value which is reached within the following three and half hours. The phase curves show negative shifts with the same dynamics. The magnitude and phase shifts recorded during the experiments were the following: ΔMagnitude = 1.16 ± 0.26 Ω, ΔPhase = -0.56 ± 0.19 °.

To explain the changes in the sensor signal, it was assumed that only the HGF mediated activation of membrane K+ currents [94] and the decrease in cell-cell contacts contribute to the observed changes in the ITO microelectrode spectrum. Higher K+ currents mean an increased outflow of positive potassium ions from the cell membrane. Since decreased cell-cell contacts should reduce the overall resistance between the electrodes by increasing the number of conductive paths through the cell layer, it was assumed that the increased magnitude is dominated by the increase of K+ currents through the cell membrane. However, the mechanism of how this would affect the sensor response is not fully understood. One possible explanation may be that the altered ion exchange through the cell membrane could modify the equilibrium between the free ions (which are assumed to be the mobile charge carriers) and the ion-pairs (which are assumed not to contribute to the transport of charge) [96] so that the resulting conductivity of the medium is reduced.

8.6. Discussion of results

Preliminary tests performed in order to characterize the response of the sensor array to different experimental parameters revealed the QCRs to be sensitive to liquid flows greater than 4 ml/h. As well, the ITO microelectrodes showed a response to acidification of medium produced by the cell metabolism. Yet, serum free medium and 10% FCS medium could not be distinguished by any of the sensors. This gave a limitation to the maximum flow rate allowed. As well, it means that the two media could be exchanged during cell cultivation experiments without triggering any sensor response which would not be generated by changes taking place in the cell layer. As a result, an optimal experimental protocol common to all experiments could be implemented along with a seeding procedure for equal partitioning of the cell suspension on each QCR of the array. Furthermore, cell experiments were carried out following a sequence intended for the characterization of the sensor response to the cell behaviour in relation to different cultivation conditions in a step by step procedure. Firstly, the sensor responses to cell proliferation and cell detachment were investigated and a statistic of the sensors signal shifts at confluency could be performed. Also, an equivalent electric model following the changes in the microelectrodes' spectra according to cell confluency could be implemented. Next, cell starvation experiments comparing flowing and static 10% FCS medium with flowing and static serum free medium were performed in order to address the buffer capacity of the microfluidic system in terms of time after stop of flow. As a result it

8. Cell culture experiments

was found that the microfluidic device provides a sufficient buffer capacity to maintain constant proliferation conditions during ~1 hour without medium exchange. Moreover, sensor signals revealed that changes in the mechanical properties of the cell layer seem to be triggered by accumulation of metabolic waste products rather than depletion of nutrients. In the following set of experiments, the optimal flow rate for cell proliferation was also found. Responses obtained at different flow rates revealed the flow of 1 ml/h being an optimal parameter for cell growth and confirmed the assumptions initially done to perform the first proliferation experiments. Thanks to its specific design, the biosensor array also allowed analysing the impact of the spatial cell distribution on the proliferation kinetics. In this frame, low cell confluency (~20%) at the beginning of the proliferation lead to a more variable response (when compared with 50% initial cell confluency) both in terms of absolute values and signal kinetics which could be good represented using a time constant based exponential behaviour. The symmetry of the device along with well defined dispensing procedures allowed to equally dispense MDCK-II cells inside each bioreactor unit. Also equal proliferation conditions were guaranteed by the symmetric design of the biosensor array. Therefore, for each single experiment dissimilarities of cell proliferation kinetics on the surface of different sensors could be directly attributed to different spatial cell distributions. This suggests that different processes may take place within the cell layer during growth depending on the initial cell configuration. At last, the response to HGF stimulation was considered. QCRs could recognise changes in adhesion strength and viscosity of the cell layer whereas the ITO electrodes showed sensitivity mainly to alteration of the ion exchange through the cellular membrane. Both responses took place within minutes from the stimulation and before anything could be observed through the microscope.

As it has just been mentioned, QCRs and ITO microelectrodes revealed to be sensitive to different properties of the cell layer. Therefore, combining the data acquired from both sensor principles provided complementary information for characterizing the cell behaviour during growth and stimulation. Due to their sensitivity to the mechanical properties of the cytoskeleton, especially in the close vicinity of the sensor surface, QCRs could follow in real-time the increase in acoustic load in a timeframe of minutes after cell seeding. This is directly linked to the development of focal contacts during initial cell attachment on the sensor surface. During the following few hours cell division was not taking place. However, successive shifts in resonant frequency and conductance maximum could be attributed to increased cell size through inspection of microscopy pictures. Once cell proliferation started no changes from the previous kinetics could be observed. Therefore, it was concluded that although the QCRs are sensitive to changes in the mechanical properties of the cell layer they cannot distinguish between increased cell number and augmented cell size during cell growth. Conversely, ITO microelectrodes responses did not show obvious relation to the initial cell attachment. However, they could recognise different cell behaviours during cell cultivation, namely an initial lag phase where no cell division takes place, followed by a cell proliferation stage and a final plateau phase where cell division is drastically reduced due to high confluency. As well, the ITO microelectrodes response recorded during HGF stimulation, lead to the assumption that this sensing principle may be sensitive to variation of the ion exchange through the cell membrane. However, further investigations should be done in this direction in order to clarify this theory and better characterise the sensor response.

9. CONCLUSIONS AND OUTLOOK

The design, simulation and fabrication of a new disposable microfluidic biosensor array conceived for the cultivation and online monitoring of adherent cells were presented. As well, impedance measurements on QCRs and transparent ITO microelectrodes were performed during cultivation and stimulation of MDCK-II cells. The results obtained corroborated the usability of the biosensor array as a tool for biological research. The device implemented is a biosensor array made of 16 QCRs and a microfluidic network allowing simultaneous multiple cell culture experiments in flow-through regime. In order to assume the cells lying on the different sensors being affected by the same experimental conditions, the microfluidic network was realized with a completely symmetrical structure. Therefore, the different paths leading to each sensor ought to have the same flow resistance and provide an equal flow distribution above all QCRs. During the design phase, COMSOL simulations addressing the influence of asymmetries within the microfluidic network, proved the assumption that the influence of fabrication tolerances on the flow regime could be neglected and further supported the hypothesis of identical experimental conditions above the sensors of the array. Moreover, the influence of the flow of media on the cell layer was addressed. This aspect is very important in terms of growth environment in order to permit the comparison of the responses acquired from cell layers cultivated in the microfluidic biosensor array (constant medium flow) with those obtained in standard cell culture flasks (no flow) applying techniques native of the molecular cell biology. When the two situations would be equivalent there would be the possibility to observe the same cellular responses with the different methods and compare them directly. Therefore, additional COMSOL simulations were performed for the estimation of the shear stress on the cell layer due to the flow-through regime. A value of about 13.3 µPa was here obtained. This value surely represents a neglectable quantity in comparison with the shear stress of 9 mPa applied in a previous study [56] to show the effects of shear stress on the morphological changes of endothelial cells. Moreover, preliminary experiments of cells cultivated in flasks placed on a shaker providing a comparably low shear force showed no significant differences in reference to a non-moving flask. Thus, the conclusion that the shear forces exerted on the cell layer due to the flow-through regime can be neglected. Moldflow simulations were also performed to optimize the geometries of the form for the injection molding fabrication of the disposable devices with minimum shrinkage and warpage. Once the best gate location has been found, different geometries were considered and simulated to identify the one giving the best results in terms

9. Conclusion and outlook

of deformation of the fabricated parts. As well, the optimization of the most important process parameters was carried out by simulation based on the material selected and the new devices could be successfully produced without defects.

After comparison of few different fabrication technologies gold electrodes were deposited by magnetron sputtering on bare quartz discs whereas ITO microelectrodes were obtained by wet etching of ITO coated glass wafers. A miniaturized impedance analyzer developed in-house [16] was then employed for the acquisition of impedance spectra. A dedicated sensor interface electronics with multiplexer was developed, characterized and optimized for fast switching and spectra acquisition of the 16 QCRs and the 16 ITO microelectrodes along with the selection of the relative reference impedance. The control of the multiplexer was implemented through the digital bits available on the miniaturized impedance analyzer. For the characterization of the new electronic interface a simple circuit model was developed in order to quantify the capacitive parasitic components of the PCB board. A three point calibration of the interface was then performed for the compensation of the parasitic components and to provide a correct measurement of the impedance values applied at the load. Before starting with cultivation of cells, the response of sensors and the stability of the measurement system were probed using water-glycerol mixtures, conductivity calibration solutions and the cell culture media. Finally, an equivalent electrical circuit was developed for fitting the spectra measured with the ITO microelectrodes. The so implemented measurement system allowed the observation fast cellular reaction kinetics. As a criticism it may be mentioned that having to deal with multiplexing of 32 sensors implies a significant number of switches which introduce parasitic components in the sensor electronic. Therefore, a redesign consisting on reducing the bioreactor units down to two would significantly simplify the electronic layout and therefore lower the parasitic capacitance C_g to about half its actual value. According to (48) this would double the bandwidth of the system available for the spectra acquisition and improve the signal to noise ratio. Moreover, by preserving two of the four independent bioreactor units the reduction in size would not affect the possibility of performing multiple experiments with simultaneously growing cell populations. As well, there are hints suggesting that slightly increasing the actual diameter (8 mm) of the quartz crystals here employed may reduce the negative influence of the paraffin seal in terms of stability of the sensor signal. In fact, the sensor signals acquired from an early "single unit" prototype [11] housing one 10 MHz 14 mm diameter QCR showed a less damped response with smaller oscillations. This was assumed to be related to a smaller fraction of the resonator being loaded by paraffin when mounted within the microfluidic device.

9. Conclusion and outlook

Nevertheless, results obtained from cell cultivation and stimulation experiments are promising. After the first investigations, which were regarded as a necessary premise to the following research, the analysis of the impact of cell distribution on the proliferation kinetics as well as HGF cell stimulation were performed. As well, an equivalent electrical model of the cell layer which could give a reliable fit of the ITO microelectrodes spectra measured for cell layers with a grade of confluency >30% was developed in this phase. Parallel cultivation of cells on the gold surface of the QCRs led to first observations of how the initial cell distribution can affect the growth of a confluent layer. The comparison of sensor signals showed that cell adhesion kinetics evaluated by means of changes in the acoustic load on the QCRs depend on the cell distribution and number of cells especially at low grades of confluency. Also, a minimum initial grade of confluency of about 20% is necessary for sufficiently comparable cell proliferation kinetics in between all sensors. As well, HGF stimulation results revealed the potential of the biosensor array in detecting the dynamics of cellular responses before any change could be observed at the microscope. The first response could be detected within minutes and oscillations of the sensor signals were taking place during a time window of four hours. Also the indication of a modified ion exchange through the cell membrane, influencing the response recorded from the ITO microelectrodes, starting shortly after the stimulation and recovering within four hours has been observed. No obvious connection between the grade of confluency and the short term response could be determined. During these experiments, the cellular behaviour during growth and stimulation could be analyzed online by means of the microfluidic biosensor array developed. By using two independent sensor principles was possible to deliver in real time complementary information about the same phenomenon. Moreover, the possibility to monitor multiple cell populations originated culture flask and simultaneously growing in the same environment represented a great advantage in comparison to "single unit" architectures. In fact, each single experiment performed with the biosensor array could already provide a statistic of the cellular behavior without the influence, for example, of a different passage number of the cells employed. As well, when an uncontrolled external factor may have occurred it would have affected all the populations at the same manner, providing again a reliable statistic when comparing sensor signals.

Concerning the future work, an accurate representation of the QCRs response able to relate directly with the properties of the cytoskeletal system is missing in literature. Therefore, additional effort should be done in the challenging task of developing such a model. In this, the analysis of the viscoelastic properties of the cell layer mediated by the acoustic load on the

QCR may well integrate with the use of models of cytoskeletal mechanics based on tensegrity or continuum elastic/viscoelastic models [97]. As well, the systematic and quantitative characterization of the cellular response mediated by the acoustic load on the QCRs and the transepithelial impedance measurement of the cell layer should be addressed. It is believed that the extensive use of the microfluidic biosensor array done in combination with the standard analysis techniques employed in molecular cell biology will allow linking the sensor responses to specific regulation mechanisms and reveal the kinetics involved. In fact, a distinctive difference between the read-outs provided by the biosensor array and, for example, fluorescent microscopy in combination with the inhibition of cell membrane receptors or immunostaining methods exists. The diversity lies in that the latter techniques can provide a rich spectrum of information at a single time point whereas the microfluidic biosensor array is able to continuously deliver real-time information about fast cell reaction kinetics. Moreover, the online measurement implemented by the sensor system is done in a non-invasive fashion. For instance, by exploring the response kinetics, the new device, would be able to identify meaningful time windows during which the major changes in response to stimulants are taking place. As a consequence, various western blot assays, used for the quantification of the protein content in a sample, would be applied in a more efficient and "response oriented" fashion rather than being performed at constant time intervals on different cell cultures. Insofar little about the kinetics of cell signalling is known. Therefore, it is believed that in the future, the synergy between the new biosensor array and the techniques native to molecular biology will provide deeper insight not only in compiling but also in the exploration of the kinetic behaviour of the cell signalling networks.

REFERENCES

[1] W.T. Gibson, M.C. Gibson, "Cell topology, geometry, and morphogenesis in proliferating epithelia" Curr Top Dev Biol., vol. 114, pp. 89-87, 2009

[2] Y. Churin, L. Al-Ghoul, O. Kepp, T. F. Meyer, W. Birchmeier, M. Naumann, "Helicobacter pylori CagA protein targets the c-Met receptor and enhances the motogenic response", The Journal of Cell Biology, vol. 161, pp. 249-255, 2003.

[3] R. Franke, M. Mueller, N. Wundrack, E.-D. Gilles, S. Klamt, T. Kähne, M. Naumann, "Host-pathogen systems biology: logical modelling of hepatocyte growth factor and Helicobacter pylori induced c-Met signal transduction", BioMed Central Systems Biology, vol. 2, pp. 1-17, 2008.

[4] B. Alberts, A. Johnson, J. Lewis, M. Raff, K. Roberts, P. Walter, "Molecular biology of the cell" Garland Science 2002", ISBN 0815332181

[5] A. Alessandrini, M.A. Croce, R. Tiozzo, P. Facci, "Monitoring Cell-cycle-related viscoelasticity by a quartz crystal microbalance", Applied Physics Letters, vol. 88, pp. 083905-1-3, 2006.

[6] J. Wegener, J. Seebach, A. Janshoff, H.-J. Galla, "Analysis of the Composite Response of Shear Wave Resonators to the Attachment of Mammalian Cells", Biophysical Journal, vol.78, pp.2821-2833, 2000.

[7] J. Wegener, A. Janshoff and H.-J. Galla, "Cell adhesion monitoring using a quartz crystal microbalance: comparative analysis of different mammalian cell lines", European Biophysical Journal, vol. 28, pp. 26-37, 1998.

[8] J. Wegener, A. Janshoff, C. Steinem, "The Quartz Crystal Microbalance as a Novel Means to Study Cell-Substrate Interactions in Situ", Cell Biochemistry and Biophysics, vol. 34, pp 121-151, 2001.

[9] S.J. Martin, V.E. Granstaff, G.C. Frye, "Characterization of quartz crystal microbalance with simultaneous mass and liquid loading", Analytical Chemistry, vol. 63, pp. 2272–2281, 1991.

[10] E. Nwankwo, C. J. Durning, "Impedance analysis of thickness-shear mode quartz crystal resonators in contact with linear viscoelastic media", Review of scientific instruments, vol. 69, pp. 2975-2384, 1998.

[11] T. Jacobs, A. Gomide, T. Kähne, A. Kienle, M. Naumann, P. Hauptmann, "Micro Fluidic Biosensor System Based on Quartz Crystal Resonators for Fast Online Adherent Cell Proliferation and Stimulation Analysis", Proceedings of the 6th IEEE Sensors Conference, pp. 720-723, 2007.

[12] T. Jacobs, K. Bolaeva, T. Kähne, M. Naumann, P. Hauptmann, "Real-time Analysis of Hepatocyte Growth Factor Induced Cell Motility with Quartz Crystal Resonators", Proceedings of the 7th IEEE Sensors Conference, pp 246-249, 2008.

[13] S. Gritsch, P. Nollert, F. Jähnig, E. Sackmann, "Impedance Spectroscopy of Porin and Gramicidin Pores Reconstituted into Supported Lipid Bilayers on Indium-Tin-Oxide Electrodes", Langmuir, vol. 14, pp. 3118-3125, 1998.

References

[14] A. N. Asanov, W. W. Wilson, P. B. Oldham, "Regenerable Biosensor Platform: A Total Internal Reflection Fluorescence Cell with Electrochemical Control", Anal. Chem., vol 70, pp. 1156-1163, 1998

[15] G. W. Gross, B. K. Rhoades, D. L. Reust, F. U. Schwalm, "" Stimulation of monolayer networks in culture through thin-film indium-tin oxide recording electrodes", Journal of Neuroscience Methods, vol. 50, pp. 131-143, 1993.

[16] S. Doerner, T. Schneider, P. Hauptmann, "Wideband impedance spectrum analyzer for process automation applications", Review of Scientific Instruments, vol. 78, paper 105101, 2007.

[17] H. Rees "Mold Engineering" Hanser (SPE books), 1995, ISBN 3446177299

[18] H. Belofsky "Plastics: product design and process engineering" Hanser (SPE books), 1995, ISBN 3446174176

[19] Milton Ohring "Materials Science of Thin Films – Deposition and structure" Elsevier, 2002, ISBN 9780125249751

[20] Nalwa, Hari Singh "Handbook of thin film materials – Vol.1/ Deposition and processing of thin films" Academic press, 2002, ISBN 0125129092

[21] A. Sapper "Mechanics and Dynamics of Liposomes and Cells Studied by QCM and EICS" 2006, PhD Dissertation

[22] D.A. Borkholder "Cell-based Biosensor using Microelectrodes"1998, PhD Dissertation

[23] K. Burridye, K. Fath, T. Kelly, G. Nuckolls, C. Turner "FOCAL ADHESIONS: Transmembrane Junctions between the Extracellular Matrix and the Cytoskeleton" Ann. Rev. Cell Biol, vol. 4, pp. 487-525, 1988

[24] D.E. Ingber "Cellular tensegrity: defining new rules of biological design that govern the cytoskeleton" Journal of Cell Science, vol. 104, pp. 613-627 1993

[25] B. Geiger, A. Bershadsky "Assembly and Mechanosensory function of focal contacts" Current Opinion in Cell Biology, vol. 13, pp. 584–592, 2001

[26] C. Katsaros, D. Karyophyllis, B. Galatis " Cytoskeleton and Morphogenesis in Brown Algae" Annals of Botany 97: 679–693, 2006

[27] S.K.Mitra, D.A.Hanson, D.D. Schlaepfer "Focal adhesion kinase: in command and control of cell motility" Nature reviews – Molecular cell biology, vol. 6, pp 56 – 68, 2005.

[28] C. Brakebusch, R. Fassler, "The integrin–actin connection, an eternal love affair." EMBO J., vol. 22, pp 2324–2333, 2003.

[29] E. Zamir, B. Geiger "Components of cell-matrix adhesions" J. of Cell Science, vol. 114, pp3577-3579, 2001.

References

[30] H.P. Schwan "Biological effects of non-ionizing radiations: cellular properties and interactions" Annals of Biomedical Engineering, vol. 16, pp.245-263, 1988.

[31] O.G. Martinsen, S Grimmes, H.P. Schwan "Interface phenomena and dielectric properties of biological tissue" Encyclopedia of Surface and Colloid Science, pp.2643-2652, 2002

[32] C. Polk "Handbook of biological effects of electromagnetic fields" CRC Press, 1996, ISBN 0849306418.

[33] C. Lo, J. Ferrier "Impedance analysis of fibroblastic cell layers measured by electric cell-substrate impedance sensing" Physical Review E, Vol 57, pp 6982-6987, 1998

[34] C. Lo, C.R. Keese, I: Giaever "Impedance analysis of MDCK cells measured by Electric Cell-Substrate Impedance Sensing" Biophysical Journal, vol. 69, pp. 2800-2807, 1995

[35] D. Salt, "The Hy-Q Handbook of Quartz Crystal Devices", Van Nostrand Reinhold Company, New York, 1987

[36] V.E. Bottom, "Introduction to Quartz Crystal Unit Design", Van Nostrand Reinhold electrical/computer science and engineering series, New York, 1982

[37] R. Lucklum, P.Hauptmann "The Quartz Crystal Microbalance: Mass sensitivity, Viscoeleasticity, and Acoustic Amplification" Sensors and Actuarors B, vol. 70, pp.30-36, 2000

[38] R. Lucklum, P.Hauptmann "Transduction mechanism of acoustic-wave based chemical and biochemical sensors" Meas. Sci. Technol., vol. 14 pp.1854-1864, 2003.

[39] R. Lucklum, P.Hauptmann "The Generalized Acoustic load concept for QCM – Mass sensitivity, Viscoeleasticity, and Other Phenomena" Chemical sensors, pp.41-44

[40] A. Arnau "Piezoelectric Transducers and Applications" Springer, 2004, ISBN 350209980

[41] C. Filiatre., G. Bardeche, M. Valentin. "Transmission-line model for immersed quartz-crystal sensors" Sensors and Actuators A, vol. 44, pp. 137-144, 1994

[42] J.F. Rosenbaum, "Bulk Acoustic Wave Theory and Devices", Artech House Acoustics Library, 1988

[43] U. Hempel "Lateral filed excited quartz crystal resonator – From theoretical approach to sensor application", Dissertation, Otto-von-Guericke-Universität Magdeburg, 2008

[44] C. Behling, R. Lucklum, P. Hauptmann. "Possibilities and limitations in quantitative determination of polymer shear parameters by TSM resonators", Sensors and Actuators A, vol. 61, pp. 260-266, 1997

[45] G. Sauerbrey "Verwendung von Schwingquarzen zur Wägung dünner Schichten und zur Mikrowägung", Zeitschrift für Physik, vol. 155, pp. 206-222, 1959

[46] K.K. Kanazawa, J.G. Gordon "Frequency of a quartz microbalance in contact with liquid", Anal. Chem., vol. 57, pp. 1770-1771, 1985

References

[47] K.K. Kanazawa, J.G. Gordon "The oscillation frequency of a quartz resonator in contact with a liquid" Anal. Chim. Acta, vol. 175, pp. 99-105, 1985

[48] H. Muramatsu, E. Tamiya, I. Karube "Computation of equivalent circuit parameters of quartz crystals in contact with liquids and study of liquid properties" Analytical Chemistry, vol. 60, pp. 2142-2146, 1988

[49] C. Behling, R. Lucklum, P. Hauptmann, "The non-gravimetric quartz crystal resonator response and its application for determination of polymer shear modulus". Meas. Sci. Technol., vol. 9, pp. 1886 – 1893, 1998.

[50] J.R. Macdonald "Impedance Spectroscopy. Theory, Experiment, and Applications" Wiley-Interscience, 2005, ISBN 0471647497.

[51] G. T. A. Kovacs, "Introduction to the theory, design, and modelling of thin-film microelectrodes for neural interfaces", in Enabling Technologies for cultured Neural Networks, D.A. Stenger and T. M. McKenna, Academic Press, London, pp. 121-165, 1994.

[52] A.J. Bard, L.R. Faulkner, "Electrochemical methods", John Wiley, New York, 2001, ISBN 0471043729

[53] J. Newman, "Resistance for flow of current to a disk", J. Electrochemical Society, vol. 113, pp. 501-502, 1966.

[54] J. Wegener, M. Sieber, H.-J. Galla, "Impedance analysis of epithelial and endothelial cell monolayers cultured on gold surfaces", Journal of Biochemical and Biophysical Methods, vol. 32, pp. 151-170, 1996.

[55] H.A. Praetorius, K.R. Spring, "Bending the MDCK Cell Primary Cilium Increases Intracellular Calcium", The journal of Membrane Biology, vol. 184, pp. 71–79, 2001

[56] V.D. Bhat, P.A. Windridge, R.S. Cherry, L.J. Mandel "Fluctuating shear stress effects on stress fiber architecture and energy metabolism of cultured renal cells" Biotechnology progress, vol. 11, pp. 596-600, 1995

[57] D. P. Gaver, S. M. Kute, "A Theoretical Model Study of the Influence of Fluid Stresses on a Cell Adhering to a Microchannel Wall", Biophysical Journal, vol. 75, pp 721-733, 1998.

[58] Z. Zhang, B. Jiang, "Optimal Process Design of Shrinkage and Sink Marks in Injection Molding", Journal of Wuhan University of Technology, Mater. Sci. Ed., vol. 22, pp. 404-407, 2007.

[59] H. Ming-Chih, T. Ching-Chih, "The effective factors in the warpage problem of an injection-molded part with a thin shell feature", Journal of Materials Processing Technology, vol. 110, pp. 1-9, 2001.

[60] J. Schroeder, S. Doerner, T. Schneider, P. Hauptmann, "Analogue and digital sensor interfaces for impedance spectroscopy", Meas. Sci. Technol., vol. 15, pp. 1271–1278, 2004.

[61] http://www.dow.com/glycerine/resources/physicalprop.htm.

[62] M. Scholten, J. Vandenmeerakker, "On the Mechanism of ITO Etching – the Specificity of Halogen Acids", J Electrochem Soc. vol. 140, pp. 471-475, 1993.

References

[63] M.F. Chen, Y.P. Chen, W.T. Hsiao, Z.P. Gu, "Laser direct write patterning technique of indium tin oxide film" Thin Solid Films, vol. 515, pp. 8515-8518, 2007

[64] G. Raciukaitis, M. Brikas, M. Gedvilas, T. Rakickas, "Patterning of indium–tin oxide on glass with picosecond lasers", Applied Surface Science, vol. 253, pp. 6570-6574, 2007

[65] O. Yavas, M. Takai, "Effect of substrate absorption on the efficiency of laser patterning of indium tin oxide thin films", Journal of Applied Physics, vol. 85, pp. 4207-4212, 1999

[66] J. Schroeder, "Miniaturisierter Impedanzanalysator und hochfrequente Sensorarrays für die Quarzmikrobalance in Flüssigkeiten", PhD Dissertation, Ottovon-Guericke-Universität Magdeburg, 2003

[67] HP 4395A Operation Manual, Network/Spectrum/Impedance Analyzer, 2nd Edition, Hewlett Packard, Inc., (1998)

[68] Leica Microsystems, The Abstract Hardware Model Programmer's Guide for DM[I] Series / Stereo- and Macroscopes, Version 3.1, April 2008

[69] K.W. Oh, C.H. Ahn, "A review of microvalves", Journal of Mechanics and Microengineering, vol. 16, pp. R13-R39, 2006

[70] J.Y Baek, J.Y. Park, J.I. Ju, T.S. Lee, S.H. Lee, "A pneumatically controllable flexible and polymeric microfluidic valve vabricated via in situ development", Journal of Mechanics and Microengineering, vol. 15, pp. 1015-1020, 2005

[71] E.T. Carlen, C.H. Mastrangelo, "Electrothermally activated paraffin microactuators", Journal of microelectromechanical systems, vol. 11, pp. 165-174, 2002

[72] J.S. Go, S. Shoji "A disposable, dead-volume free and leak-free in plane PDMS microvalve", Sensors and Actuators A, vol. 114, pp. 438-444, 2004.

[73] H.G. Hong, Y. Kim, "Electrochemical characteristics of an indium-tin oxide electrode modified with 2,5-bis(phosphomethyl)hydroquinone", Electrochimica Acta, vol. 46, pp. 2313-2319, 2001.

[74] X.B. Yang, H.I. Roach, N.M. Clarke, S.M. Howdle, R. Quirk, K.M. Shakesheff, R.O.C. Oreffo, "Human osteoprogenitor growth and differentiation on synthetic biodegradable structures after surface modification" Bone, vol. 29, pp. 523–531, 2001

[75] T.J. Webster, C. Ergun, R.H. Doremus, R.W. Siegel, R. Bizios, "Enhanced functions of osteoblasts on nanophase ceramics." Biomaterials, vol.21, pp. 1803–1810, 2000

[76] M.C. Serrano, R. Pagani, M. Vallet-Regi, J. Pena, A. Ramila, I. Izquierdo, M.T. Portoles, "In vitro biocompatibility assessment of poly(epsiloncaprolactone) films using L929 mouse fibroblasts". Biomaterials, vol. 25, pp.5603–5611, 2004

[77] A.K. Shah, R.K. Sinha, N.J. Hickok, R.S. Tuan, "High-resolution morphometric analysis of human osteoblastic cell adhesion on clinically relevant orthopedic alloys." Bone, vol. 24, pp. 499-506, 1999

References

[78] B.G. Keselowsky, D.M. Collard, A.J. Garcia, "Surface chemistry modulates fibronectin conformation and directs integrin binding and specificity to control cell adhesion", J Biomed Mater Res A, vol. 66, pp. 247-259, 2003.

[79] A. Rezania, K.E. Healy, "Biomimetic peptide surfaces that regulate adhesion, spreading, cytoskeletal organization, and mineralization of the matrix deposited by osteoblast-like cells", Biotechnol Prog, vol. 15, pp. 19-32, 1999

[80] S.P. Wolsky, A.W. Czanderna, "Methods and phenomena 7 – Application of piezoelectric quartz crystal microbalances", Elsevier, 1991, ISBN 0444422773

[81] R. Lucklum, P. Hauptmann, "Determination of polymer shear modulus with quartz crystal resonators, Faraday Discuss., vol. 107, pp. 123 - 140, 1997

[82] R. Lucklum, B. Henning, P. Huptmann, D. Schierbaum, S. Vahiniger, W. Göpel, "Quartz Microbalance Sensors for Gas Detection", Sensors and Actuators A, vol. 27, pp. 705-710, 1991

[83] K.A. Marx, T. Zhou, M. Warren, S.J. Braunhut, "Quartz crystal microbalance study of endothelial cell number dependent differences in initial adhesion and steady-state behavior: evidence for cell–cell cooperativity in initial adhesion and spreading" Biotechnol Prog, vol. 19, pp. 987 999, 2003

[84] C.M. Marxer, M.C. Coen, T. Greber, U.F. Greber, L. Schlapbach, "Cell spreading on quartz crystal microbalance elicits positive frequency shifts indicative of viscosity changes" Anal Bioanal Chem, vol. 377, pp. 578-586, 2003

[85] K.A. Marx, T.A. Zhou, A. Montrone, H. Schulze, S.J. Braunhut, "A quartz crystal microbalance cell biosensor: detection of microtubule alterations in living cells at nM nocodazole concentrations", Biosens Bioelectron, vol. 16, pp 773-782, 2001

[86] C. Fredriksson, S. Kihlman, M. Rodahl, B. Kasemo, "The piezoelectric quartz crystal mass and dissipation sensor: a means of studying cell adhesion", Langmuir vol. 14, pp. 258-251, 1998

[87] M. Rodahl, F. Hook, C. Fredriksson, C.A. Keller, A. Krozer, P. Brzezinski, "Simultaneous frequency and dissipation factor QCM measurements of biomolecular adsorption and cell adhesion", Faraday Discuss, vol. 107, pp. 229-246, 1997

[88] I. Royal, N. Lamarche-Vane, L. Lamorte, K. Kaibuchi, M. Park, "Activation of Cdc42, Rac, PAK, and Rho- Kinase in Response to Hepatocyte Growth Factor Differentially Regulates Epithelial Cell Colony Spreading and Dissociation", Molecular Biology of the Cell, vol. 11, pp. 1709-1725, 2000.

[89] K. Takaishi, T. Sasaki, M. Kato, W. Yamochi, S. Kuroda, T. Nakamura, M. Takeichi, Y. Takai, "Involvement of Rho p21 small GTP-binding protein and its regulator in the HGF-induced cell motility", Oncogene, vol. 9, pp. 273-279, 1994

[90] C. D. Nobes, A. Hall, "Rho GTPases Control Polarity, Protrusion and Adhesion during Cell movement", The journal of Cell Biology, vol. 144, pp. 1235-1244, 1999.

References

[91] T. Nakamura, T. Nishizawa, M. Hagiya, T. Seki, M. Shimonishi, A. Sugimura, K. Tashiro, S, Shimizu, "Molecular cloning and expression of human hepatocyte growth factor", Nature, vol 342, pp. 440 – 443, 1989.

[92] L. Naldini, K.M. Weidner, E. Vigna, G. Gaudino, A. Bardelli, C. Ponzetto, R.P. Narsimhan, G. Hartmann, R. Zarnegar, G.K. Michalopoulos, W. Birchmeierl, P.M. Comoglio, "Scatter factor and hepatocyte growth factor are indistinguishable ligands for the MET receptor", The EMBO Journal, vol.10, pp.2867-2878, 1991

[93] D. Tulasne, B. Foveau "The shadow of death on the MET tyrosine kinase receptor", Cell Death and Differentiation, vol.15, pp. 427-434, 2008.

[94] M. Jin, D.M. Defoe, R. Wondergem, Hepatocyte Growth Factor/Scatter Factor Stimulates Ca2+-Activated Membrane K+ Current and Migration of MDCK II Cells, J. Membrane Biology, vol. 191, pp. 77–86, 2002

[95] J.M. Arthur, "The MDCK cell line is made up of populations of cells with diverse resistive and transport properties", Tissue and Cell, vol. 32, pp. 446-450, 2000.

[96] F.H. Stillinger, "Ion distribution in concentrated electrolytes" Proc. Natl. Acad. Sci. U S A, vol. 60, pp. 1138-1143, 1968.

[97] M.R.K. Mofrad, R.D. KAMM, "Cytoskeletal Mechanics – Models and Measurement", Cambridge University press, 2006, ISBN 978052184637

Die VDM Verlagsservicegesellschaft sucht für wissenschaftliche Verlage abgeschlossene und herausragende

Dissertationen, Habilitationen, Diplomarbeiten, Master Theses, Magisterarbeiten usw.

für die kostenlose Publikation als Fachbuch.

Sie verfügen über eine Arbeit, die hohen inhaltlichen und formalen Ansprüchen genügt, und haben Interesse an einer honorarvergüteten Publikation?

Dann senden Sie bitte erste Informationen über sich und Ihre Arbeit per Email an *info@vdm-vsg.de*.

Sie erhalten kurzfristig unser Feedback!

VDM Verlagsservicegesellschaft mbH
Dudweiler Landstr. 99
D - 66123 Saarbrücken
Telefon +49 681 3720 174
Fax +49 681 3720 1749

www.vdm-vsg.de

Die VDM Verlagsservicegesellschaft mbH vertritt

Printed by Books on Demand GmbH, Norderstedt / Germany